Particles Separation in Microfluidic Devices, Volume II

Particles Separation in Microfluidic Devices, Volume II

Editors

Takasi Nisisako
Naotomo Tottori

MDPI • Basel • Beijing • Wuhan • Barcelona • Belgrade • Manchester • Tokyo • Cluj • Tianjin

Editors
Takasi Nisisako
Institute of Innovative Research
Tokyo Institute of Technology
Yokohama
Japan

Naotomo Tottori
Department of Mechanical
Engineering
Kyushu University
Fukuoka
Japan

Editorial Office
MDPI
St. Alban-Anlage 66
4052 Basel, Switzerland

This is a reprint of articles from the Special Issue published online in the open access journal *Micromachines* (ISSN 2072-666X) (available at: www.mdpi.com/journal/micromachines/special_issues/Particles_Separation_Volume_II).

For citation purposes, cite each article independently as indicated on the article page online and as indicated below:

LastName, A.A.; LastName, B.B.; LastName, C.C. Article Title. *Journal Name* **Year**, *Volume Number*, Page Range.

ISBN 978-3-0365-3674-3 (Hbk)
ISBN 978-3-0365-3673-6 (PDF)

Contents

About the Editors

Takasi Nisisako

Takasi Nisisako received a Ph.D. in mechanical engineering from the University of Tokyo, Japan, in 2005. Presently, he is an associate professor working at the Institute of Innovative Research, Tokyo Institute of Technology, Japan. His current research interests mainly include microfluidic systems for analytical and/or manufacturing applications.

Naotomo Tottori

Naotomo Tottori received a Ph.D in mechanical engineering from the Tokyo Institute of Technology in 2018. In 2018–2020, he was a specially appointed assistant professor working at the Institute of Innovative Research, Tokyo Institute of Technology, Japan. Presently, he is working as an assistant professor with the Department of Mechanical Engineering, Faculty of Engineering, Kyushu University, Japan. His current research interests mainly include microfluidic systems for particles processing.

micromachines

Editorial

Editorial for the Special Issue on Particles Separation in Microfluidic Devices, Volume II

Naotomo Tottori [1,*] and Takasi Nisisako [2,*]

1 Department of Mechanical Engineering, Faculty of Engineering, Kyushu University, W4-729, 744, Motooka, Nishi-ku, Fukuoka 819-0395, Japan
2 Institute of Innovative Research, Tokyo Institute of Technology, R2-9, 4259 Nagatsuta-cho, Midori-ku, Yokohama 226-8503, Japan
* Correspondence: tottori@mech.kyushu-u.ac.jp (N.T.); nisisako.t.aa@m.titech.ac.jp (T.N.)

Citation: Tottori, N.; Nisisako, T. Editorial for the Special Issue on Particles Separation in Microfluidic Devices, Volume II. *Micromachines* **2022**, *13*, 482. https://doi.org/10.3390/mi13030482

Received: 17 March 2022
Accepted: 17 March 2022
Published: 20 March 2022

Publisher's Note: MDPI stays neutral with regard to jurisdictional claims in published maps and institutional affiliations.

Particle separation in the nano- to microscale range is a significant step for biological, chemical, and medical analyses. Since the 2000s, many micro- and nanofluidic techniques have been developed for the separation, sorting, isolation, fractionation, and purification of various particles, especially for biological particles, based on their chemical and/or physical properties. These micro- and nanofluidic techniques have been attracting attention as a promising approach because they enable us to use a sample and reagent with minimum consumption, are easy to use, and can be integrated into other components for comprehensive analysis. These micro- and nanofluidic techniques are classified into two types; (i) passive techniques using the hydrodynamic effect induced by the geometries in the micro/nanoscale, and (ii) active techniques using external forces, for example in the magnetic, optical, acoustic, and electric fields.

The papers collected in this Special Issue present state-of-the-art research for particle separation and manipulation using microfluidic techniques. There are five research papers and a review article published in this Special Issue. Three research papers present the separation and manipulation of particles by measn of the passive approach using layers of stacked beads in a microchannel [1], three-dimensional (3D) channels [2], and convex air bubbles attached to the surface of the channel [3]. The remaining papers present the separation of particles by means of the integration of active and passive techniques using an electric field and a deterministic lateral displacement (DLD) array [4,5]. The review article covers microfluidic technologies for separation and detection toward achieving practical applications for public health [6].

(1) Passive method: Xu et al. proposed a microfluidic plasma separator for biochemical analysis using three layers of different-sized microspheres as the separation structure [1]. They designed a microfluidic device with 18 capillary microchannels and enabled the extraction of ~3 µL of plasma from a 50 µL blood sample in ~55 min. As a demonstration of the feasibility of the device in the application of clinical biochemical testing, they introduced a clinical blood sample into the device and measured the concentrations of four components (TP, ALB, GLU, and UA), and obtained measurement values similar to those provided by conventional centrifugal separation, indicating acceptable accuracy for point-of-care analysis. Zoupanou et al. reported a method for the fabrication of a 3D fluidic mixer made from poly(methylmethacrylate) (PMMA) using computer-aided design (CAD) and the micromilling technique and demonstrated the manipulation of the fluid and nanoparticles [2]. They performed passive chaotic mixing and dilution through the reservoir and serpentine layers in the fabricated device. Park et al. proposed a novel technique to prevent the clogging of microspheres in the channel of the catheter by utilizing convex air bubbles attached to the surface of the channel walls [3]. In this paper, they evaluated the effect of the width, cavity, and the distances between the cavities to prevent the clogging of microspheres.

From the experimental results, they established that the large convex air bubbles with a small distance between the two adjacent cavities can effectively prevent the clogging of microspheres in the channel.

(2) Integration of the passive and active methods: Ho et al. reported the separation of particles with nano- and micro-sizes primarily based on their zeta potential by combining DLD methods with electric fields (eDLD) [4]. In this study, they performed the characterization of the relevant parameters (e.g., ionic strength, applied voltage, frequency, and pressure) necessary to achieve adequate separation of particles with different types, enabling them to adapt the method to various particle sizes and zeta potentials. They also demonstrated the separation of nano-sized liposomes with different lipid components with biological relevance. Ho et al. also reported the separation of cells based on their differences in the membrane and/or internal structure by combining a DLD technique with electrokinetics [5]. Using the proposed microfluidic device, they performed the separation of cells, which were heat-treated to deactivate cells for changing their viability and structure, based on their viability (viable or non-viable cells). For the separation of *Escherichia coli* (*E. coli*), the change in their size after deactivation is not sufficient for the size-based separation of the DLD array; however, they utilized the considerable change in their zeta potential, and achieved separation of *E. coli* by applying AC voltage in the DLD array at a low frequency. In contrast, for the separation of *Saccharomyces cerevisiae* (Baker's yeast), the change in zeta potential after heat treatment is small, and therefore they utilized the change in dielectrophoretic property and achieved the separation of Baker's yeast by applying AC voltage in the DLD array at a higher frequency.

In addition to these research articles, Zhang et al. presented a comprehensive review of the latest developments of microfluidic separation and detection for the benefits of public health. These two topics (i.e., separation and detection) are normally reviewed separately; however, they are related closely to each other for their application to public health, and understanding the techniques of separation will offer new insights to develop detection technologies.

We would like to thank all authors who submitted their papers to this Special Issue. We would also like to acknowledge all of the reviewers who dedicated their time to provide their reports in a timely manner in order to improve the quality of the papers collated in this Special Issue.

Conflicts of Interest: The authors declare no conflict of interest.

References

1. Xu, H.; Wu, Z.; Deng, J.; Qiu, J.; Hu, N.; Gao, L.; Yang, J. Microsphere-Based Microfluidic Device for Plasma Separation and Potential Biochemistry Analysis Applications. *Micromachines* **2021**, *12*, 487. [CrossRef] [PubMed]
2. Zoupanou, S.; Chiriacò, M.S.; Tarantini, I.; Ferrara, F. Innovative 3D Microfluidic Tools for On-Chip Fluids and Particles Manipulation: From Design to Experimental Validation. *Micromachines* **2021**, *12*, 104. [CrossRef] [PubMed]
3. Park, D.H.; Jung, Y.J.; Steve Jeo Kins, S.J.K.; Kim, Y.D.; Go, J.S. Prevention of Microsphere Blockage in Catheter Tubes Using Convex Air Bubbles. *Micromachines* **2020**, *11*, 1040. [CrossRef] [PubMed]
4. Ho, B.D.; Beech, J.P.; Tegenfeldt, J.O. Charge-Based Separation of Micro- and Nanoparticles. *Micromachines* **2020**, *11*, 1014. [CrossRef] [PubMed]
5. Ho, B.D.; Beech, J.P.; Tegenfeldt, J.O. Cell Sorting Using Electrokinetic Deterministic Lateral Displacement. *Micromachines* **2021**, *12*, 30. [CrossRef] [PubMed]
6. Zhang, X.; Xu, X.; Wang, J.; Wang, C.; Yan, Y.; Wu, A.; Ren, Y. Public-Health-Driven Microfluidic Technologies: From Separation to Detection. *Micromachines* **2021**, *12*, 391. [CrossRef] [PubMed]

Article

Microsphere-Based Microfluidic Device for Plasma Separation and Potential Biochemistry Analysis Applications

Hongyan Xu [1], Zhangying Wu [1], Jinan Deng [1], Jun Qiu [2,*], Ning Hu [1], Lihong Gao [3,*] and Jun Yang [1,*]

1. Key Laboratory of Biorheological Science and Technology, Ministry of Education and Bioengineering College, Chongqing University, Chongqing 400030, China; 201819021046@cqu.edu.cn (H.X.); 202019131112@cqu.edu.cn (Z.W.); biojdeng@cqu.edu.cn (J.D.); huning@cqu.edu.cn (N.H.)
2. Department of Information, First Affiliated Hospital, Army Medical University, Chongqing 400042, China
3. Chongqing Center for Drug Evaluation and Certification, Chongqing 401120, China
* Correspondence: xnyyxxk@tmmu.edu.cn (J.Q.); glh19712@outlook.com (L.G.); bioyangjun@cqu.edu.cn (J.Y.); Tel.: +86-23-6875-4443 (J.Q.); +86-23-6035-3856 (L.G.); +86-23-6510-2291 (J.Y.)

Abstract: The development of a simple, portable, and cost-effective plasma separation platform for blood biochemical analysis is of great interest in clinical diagnostics. We represent a plasma separation microfluidic device using microspheres with different sizes as the separation barrier. This plasma separation device, with 18 capillary microchannels, can extract about 3 µL of plasma from a 50 µL blood sample in about 55 min. The effects of evaporation and the microsphere barrier on the plasma biochemical analysis results were studied. Correction factors were applied to compensate for these two effects. The feasibility of the device in plasma biochemical analysis was validated with clinical blood samples.

Keywords: microchip; microspheres stacking; plasma separation; concentration detection

Citation: Xu, H.; Wu, Z.; Deng, J.; Qiu, J.; Hu, N.; Gao, L.; Yang, J. Microsphere-Based Microfluidic Device for Plasma Separation and Potential Biochemistry Analysis Applications. *Micromachines* **2021**, *12*, 487. https://doi.org/10.3390/mi12050487

Academic Editors: Takasi Nisisako and Naotomo Tottori

Received: 30 March 2021
Accepted: 21 April 2021
Published: 26 April 2021

Publisher's Note: MDPI stays neutral with regard to jurisdictional claims in published maps and institutional affiliations.

1. Introduction

Blood biochemical tests are widely adopted for the screening of diseases in clinical diagnosis. However, red blood cells (RBCs) in the whole blood, with an intense red color, can interfere with test results [1,2]. Separation of the plasma from the whole blood sample is usually prerequired for an accurate determination of the components in a blood sample [1,3]. The centrifugation technique is a common method for plasma separation in a clinical test. However, this technique requires bulky and expensive equipment, limiting its applications in resource-limited settings [4]. Therefore, the development of a cost-effective point-of-care testing (POCT) device for plasma separation is highly desired.

Microfluidic techniques have shown great potential in the fabrication of POCT devices due to their low-cost, small size, low sample consumption and high-throughput analysis [5,6]. Numerous efforts have been made to design and fabricate microfluidic devices for plasma separation [3,7]. For example, paper-based microfluidic devices use porous cellulose material to remove RBCs from a whole blood sample [2,8–10]. Although these devices are simple and low-cost, the porous structure of the cellulose material is easily clogged by blood cells and retains proteins within its network structure [3]. Microstructures have been fabricated to perform plasma separation based on different mechanisms, such as digital microfluidics [11], cross-flow pillars [12,13], inertial force-based spiral channels [14–17], and the Zweifach–Fung separation technique [18,19]. With the elaborate design of microstructures, these devices can achieve a fast separation rate. However, the fabrication process of these microstructures is time-consuming, and precise control of the microfluid is also required for high separation efficiency. Previous research has shown that a separation system constructed by stacking microspheres in the microchannel can efficiently block blood cells and drive plasma forward by capillary force at the same time [20–23]. This method provides a simple way to selectively extract plasma from a whole blood sample without an external driving force. However, the separated plasma were directly applied in an immunoassay [22] and an agglutination test [23] assumed that the main components were perfectly preserved. The component concentrations in the plasma separated by those microfluidic devices have rarely been analyzed. It has been shown that the microstructures and surface properties of microfluidic devices can interact with the components in the plasma, which can lead to the component concentrations in the separated plasma being inconsistent with those of the "gold standard" of the centrifuge method [24,25].

In this work, we designed a microfluidic plasma separator with double-layer structure containing three types of microsphere layers. The formed microsphere barrier not only blocks RBCs but also allows the plasma to pass through by the capillary force without external driving forces. By controlling the stacking behavior of the microspheres and increasing the number of capillary channels connecting the microsphere barrier and the collection chamber, separation efficiency can be increased. The effects of water evaporation in the sample, induced by the open inlet and outlet of the device, and the microsphere barrier on the concentration variations of four components in the separated plasma were studied. Correction factors were also calculated for our device for the measurement of the four components in the separated plasma to make the results comparable to those with the traditional centrifuge method. Finally, the feasibility of using our device to separate clinical blood samples for a clinical biochemical test was studied.

2. Materials and Methods

2.1. Reagents and Materials

Phosphate buffered saline (PBS) was purchased from Solarbio Technology (Beijing, China). Protein blocking powder was obtained from Boster Biological Technology (Pleasanton, CA, USA). Photoresist (SU8 3050, SU8 2050) and the developer were provided by MicroChem (Newton, MA, USA). Polydimethylsiloxane (PDMS, Sylgard 184) was from Dow Corning (Midland, MI, USA). Microspheres with diameters of 10 μm, 20 μm and 100 μm were offered by Zhiyi Microsphere Technology (Suzhou, China).

2.2. Microchip Design and Fabrication

The microfluidic plasma separator mainly contains four parts: the injection port, the microsphere stacking chamber, the straight capillary channels, and the collection chamber (Figure 1a,b). Microspheres were stacked in the microsphere stack chamber to block the RBCs and allow the plasma to pass through. The length and height of the microsphere stack chamber are 18.5 mm and 220 μm, respectively. The maximum width of the microsphere stacking chamber is 7.8 mm. There are 18 straight capillary channels connecting the microsphere stacking chamber and the collection chamber, which can provide a capillary force to promote the separated plasma toward the collection chamber with high throughput. The dimensions of each channel are 1.5 mm (l) $\times$ 100 μm (w) $\times$ 85 μm (h). The collection chamber is designed to collect the separated plasma, with the dimensions of 5.8 mm (l) $\times$ 2 mm (w) $\times$ 220 μm (h).

Because the height of the straight capillary channel is different from other parts in the device (Figure 1b), a double-layered master mold was made on a silicon wafer by the lithography technique. SU-8 3050 and SU-8 2050 were used to prepare the first layer with the height of 85 μm, and the second layer with the height of 220 μm, respectively. To prepare the PDMS pattern, the PDMS prepolymer and the curing agent were mixed at the weight ratio of 10:1 and degassed in the vacuum oven for 30 min to remove the air bubbles. The PDMS mixture was then poured on the SU-8 master mold and cured in an oven at 80 °C for 30 min. After curing, the PDMS pattern was peeled off from the master mold and the inlets and outlets were then punched. Then, the PDMS pattern was bonded to a pre-cleaned glass slide through oxygen plasma treatment (PDC-MG, Suzhou, China). The resulting plasma separation microchip is shown in Figure 1c.

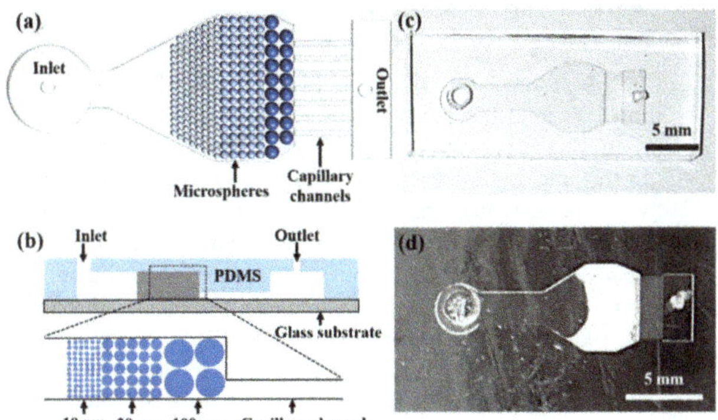

Figure 1. Schematic illustrations of the microfluidic chip of (**a**) the top view and (**b**) the cross section, and photos of the microchip (**c**) before and (**d**) after stacking microspheres.

2.3. Beads Stacking

A syringe pump (LSP10-1B, Lange, Baoding, China) was used to drive the microspheres into the stacking chamber for bead stacking. Protein blocking solutions containing 0.5 mg/mL silica microspheres with the diameter of 100 µm was first pumped into the microsphere stacking chamber at a flow rate of 600 µL/min. After all the microspheres of 100 µm were stacked at the entrance of the straight capillary tubes, the 20 µm and 10 µm microspheres were sequentially pumped into the microsphere stacking chamber at a flow rate of 600 µL/min. The volumes of 100 µm, 20 µm and 10 µm silica microspheres pumped into the chamber were 5 µL, 20 µL and 20 µL, respectively. After that, the chip was put on a hot plate at 75 °C for an hour to remove moisture, and was then left in a refrigerator (−20 °C) overnight to form the microsphere barrier to block the RBCs. The plasma separator containing the microsphere barrier is shown in Figure 1d. The length of the microsphere barrier is about 4.8 mm.

2.4. Plasma Separation

Human whole blood samples were collected from volunteers in Chongqing University Hospital. The blood samples were used for the experiment with the consent of the volunteers and Chongqing University Hospital. The blood samples used in the experiments were obtained from healthy volunteers aged between 18 and 40. Venous blood was used within 1 week after sampling and was stored in ethylenediaminetetraacetic acid (EDTA)-coated blood collection tubes at 4 °C. For the plasma separation, 50 µL blood sample was dropped into the inlet port and the plasma separation process was monitored under an optical microscope (IX73, Olympus, Tokyo, Japan) and recorded by a camera (DS126282, Canon, Tokyo, Japan).

2.5. Plasma Analysis

Assay kits for total protein (TP), albumin (ALB), glucose (GLU), and uric acid (UA), provided by Nanjing Institute of Biological Engineering (Nanjing, China), were used to measure the concentrations of components in the plasma samples. The component concentrations in the separated plasma were determined by the microplate reader (Bio Tek, Winooski, VT, USA). Plasma sample processing was performed according to the instruction by the supplier. For the measurement, 150 µL of the mixture of the plasma sample and the reaction reagent were pipetted into 96-well plate. The absorbance measurements were carried out at the corresponding wavelength with the microplate reader (Bio Tek, Winooski, VT, USA). For comparison, the concentrations of components in plasma samples prepared by the common centrifuge technique were assessed.

3. Results and Discussion

3.1. Microspheres Stacking

The major blood cells in blood are RBCs. Removing RBCs from the whole blood is important for plasma separation. The plate-shaped RBCs are 2 μm thick with a diameter of 8 μm [7]. To effectively block the RBCs and allow the plasma to pass through, the pore size of the stacking microspheres is a critical factor. Pores of too large size cannot block RBCs, while pores which are too small would be clogged by RBCs to prevent the movement of the plasma. Previous research has shown that the size of the microspheres should be controlled between 20 μm and 7.8 μm in order to prevent the movement of RBCs and allow the plasma to pass through at the same time [20]. However, if too few microspheres are stacked for the plasma separation, the amount of plasma separation collected is limited [21]. In order to achieve high-throughput plasma separation, three sizes of microspheres were used to create the barrier in this study. Large microspheres with a diameter of 100 μm are used first to prevent smaller microspheres from entering the separation channel, and which have limited contribution to the entire plasma separation process [20,23]. Plasma separation is realized by the 10 μm microspheres. However, we found that the small 10 μm microsphere could penetrate the 100 μm microsphere layer and move into the capillary channels if we directly pumped the 10 μm microsphere into the chamber, followed by the stacking of the 100 μm microsphere layer (Figure 2a). Therefore, we chose microspheres with the diameter of 20 μm as the transition layer between the two types of microspheres. After being dried on a hot plate, the microchip containing microsphere layers with clear boundaries is shown in Figure 2b.

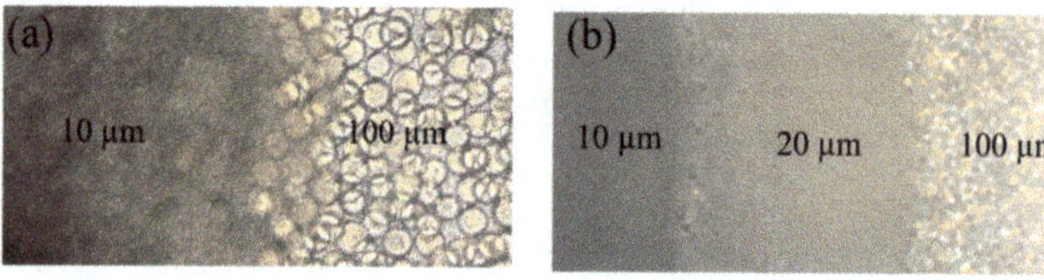

Figure 2. Optical microscopy images of (**a**) the stacking of 100 μm and 10 μm microspheres, and (**b**) stacking of microspheres with the size of 100 μm, 20 μm and 10 μm.

3.2. Plasma Separation

For the plasma separation, a 50 μL blood sample was dropped into the inlet port, allowing it to move into the microsphere stacking chamber for the plasma separation (Figure 3a). After the addition of the blood sample, the empty part of the microsphere stacking chamber quickly turned red in about 40 s, indicating that the blood sample can easily flow into the chamber without external forces. When it reached the microsphere barrier, the moving rate of the blood sample was reduced due to the large resistance of the barrier. After a while, a transparent band started to appear within the microsphere layers, which indicated that the plasma began to be separated from the blood sample (Figure 3b). The transparent plasma band gradually became clearer and enlarged (Figure 3c). After passing through the microsphere barrier, the transparent plasma quickly passed through the 18 capillary microchannels (Figure 3d, Video S1), and finally, accumulated in the collection chamber. The plasma in the capillary microchannels is transparent, suggesting that the capillary force added by the microsphere layers did not break the cell membrane of RBCs. The pores formed by the closely packed microspheres of 10 μm are too small to allow the RBCs to pass through the microsphere barrier, while the capillary force produced by the pores can promote the movement of plasma toward the collection chamber. In this way, the plasma can be separated from the blood sample. About 3 μL plasma was collected in 55 min with our device, which is acceptable for a POCT analysis [1,26,27]. Compared with the microbead-based device reported previously [21], which can only separate 350 nL plasma, the device in this study, with 18 capillary microchannels, shows a higher yield in plasma collection.

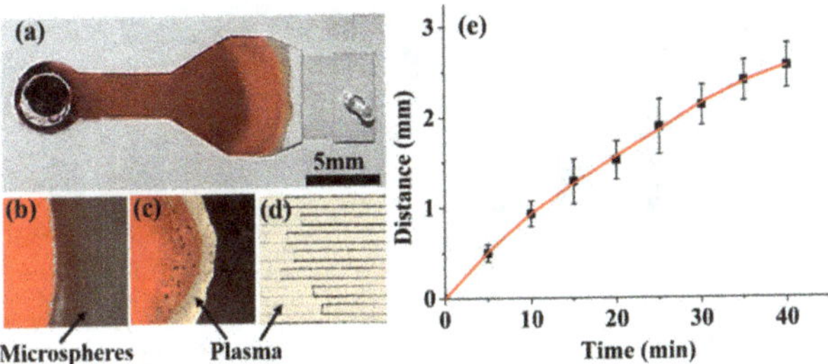

Figure 3. (**a**) Photo of the microchip during the plasma separation. Optical microscopy images of (**b**) the start separation of the plasma, (**c**) movement of the plasma within the microsphere barrier, and (**d**) flow of the plasma in the capillary microchannels. (**e**) The moving distance of the plasma as a function of time.

The movement of the front line of the separated plasma was recorded to study the kinetics of plasma separation. The zero point was set as the point where the plasma started to be separated from the blood sample. Figure 3e shows the moving distance of the midpoint of the front line as a function of time. The moving distance of the front line increases with the increase in time. The slope of the plot represents the moving rate of the separated plasma. The moving rate of the separated plasma gradually decreases with the increase of time. It took about 40 min for the plasma to pass through the microsphere barrier and the capillary channels. As the amount of the plasma separated increases, the blood cells accumulated near the edge of the microsphere barrier will increase the viscosity of the remaining blood. Therefore, the moving rate of the plasma gradually slows down, which is also reported by previous studies [7,23,28]. About 3 µL plasma can be collected in 55 min with this device.

3.3. Plasma Analysis

Proteins and various small molecules in the blood plasma are necessary for sustaining health, and they are usually adopted as biomarkers for the clinical diagnosis of diseases [1,3]. For example, glucose concentrations in the plasma are usually considered as the "gold standard" in the clinical diagnosis of diabetes [29,30]. In order to prove the feasibility of our device in the application of plasma analysis, we employed four assay kits to assess the component concentrations in the separated plasma by our device. The component concentrations in the plasma prepared by traditional centrifuge technique were used as comparison. Figure 4a shows concentrations of four components—TP, ALB, GLU, and UA—in the separated plasma by our device, and centrifugal plasma by centrifuging. The concentrations of the four components in the separated plasma show different degrees of increase compared with those in the centrifugal plasma, which show about 19.133 g/L increase for TP, 4.85 g/L increase for ALB, 0.0489 g/L increase for GLU, and 0.004 g/L increase for UA, respectively (Figure 4a).

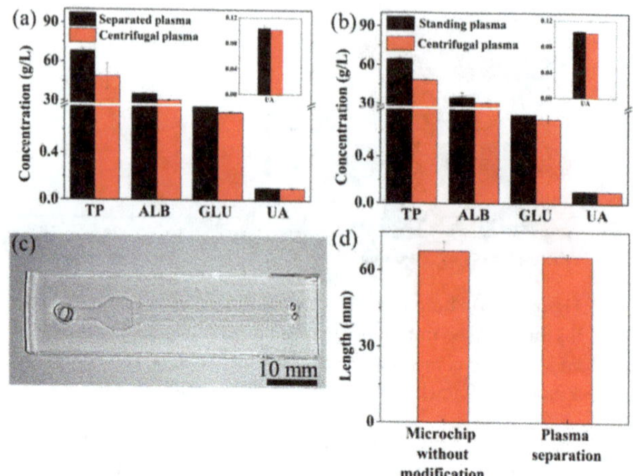

Figure 4. (**a**) Component concentrations of the separated and centrifugal plasma samples; inset is the UA concentration of separated and centrifugal plasma samples. (**b**) Absorbance of standing and centrifugal plasma; inset is the UA concentration of separated and centrifugal plasma samples. (**c**) The photo of the microdevice for the study of the effect of microsphere barrier on the measurement results. (**d**) The length of water separated from the device.

Considering the separation time (~55 min) and the dried protein blocking coating on the surface of the microspheres, we hypothesize that the evaporation of water in the sample from the open inlet and outlet of the device, and the microsphere barrier, are the main reasons that lead to the increased component concentration for the separated plasma. Standard plasma samples were used to prove our hypothesis. For the study of the influence of evaporation, a 50 µL centrifugal plasma sample was dropped into the inlet of the device and collected from the collection chamber after 55 min, before the absorbance measurement, which is the same length of time for the plasma separation process. Compared with centrifugal plasma without standing, the standing plasma shows a stronger absorbance for the four components (Figure 4b) which lead to concentration increases of 15.95 ± 0.92 g/L for TP, 4.23 ± 2.9 g/L for ALB, 0.038 ± 0.007 g/L for GLU, and $0.002 \pm 3.8 \times 10{-5}$ g/L for UA, respectively. This suggests that evaporation of the water in the plasma during the separation process can lead to an increased concentration of the components in the plasma. Then, we designed two straight channels with a scale, instead of the collection chamber connecting the capillary channels (Figure 4c), to study the influence of the microsphere barrier on the measurement results. The straight channel is 2 mm wide, 0.22 mm high and 37 mm long. The straight channel allows us to precisely measure the volume of the sample separated from the microsphere barrier. Deionized water was used to perform this experiment. After the addition of 50 µL deionized water into the device for about 10 min, the length of water going out from our device was 65.3 mm, while it was 67.4 mm for the device containing the barrier without the protein blocking coating (Figure 4d). The stacked silica microspheres in the chamber created a large surface area for the sample to make contact with. The dried hydrophilic protein blocking layer coated on the surface of microspheres can uptake some amount of water in the sample, which can lead to increased measurement results of plasma components. The volume difference between these two devices is 0.924 µL, which can lead to concentration increases of 0.919 g/L for TP, 0.571 g/L for ALB, 0.0129 g/L for GLU, and 0.0018 g/L for UA, respectively. Therefore, the effects of evaporation and microsphere barrier lead to concentration increases of 16.87 g/L for TP, 4.801 g/L for ALB, 0.051 g/L for GLU, and 0.0039 g/L for UA, which are similar to the concentration differences between the separated and centrifugal plasma samples. The slight difference is due to the different affinities between the protein blocking coating and the components, which lead to different amounts of component adhered within the microsphere barrier. In order to obtain measurement results that were comparable to those in the centrifugal sample, correction factors were calculated for the four components measured in this study (Table 1).

Table 1. Concentration correction value for the plasma component measurement.

Types of Components in Plasma	Correction Factor (g/L)	Standard Deviation
TP	19.133	8.58
ALB	4.854	0.199
GLU	0.049	0.004
UA	0.004	0.001

3.4. Clinical Validation

In order to assess the validity of the correction factors, we used blood samples from three volunteers to measure the component concentrations in the plasma separated by our device. In the process of plasma separation of each human blood sample, no hemolysis occurred. The four component concentrations in all three separated samples are higher than those in centrifugal samples before correction (Figure 5). The concentration differences for TP, ALB, GLU, and UA between the separated and the centrifugal samples are 21.203 $\pm$ 2.205 g/L, 4.978 $\pm$ 0.765 g/L, 0.043 $\pm$ 0.012 g/L, and 0.005 $\pm$ 0.001 g/L respectively. After correction, the component concentrations in separated sample show an obvious decrease and move toward the concentrations of centrifugal ones. The concentration differences of components between the corrected and centrifugal samples are 1.408 $\pm$ 1.056 g/L, 0.168 $\pm$ 0.765 g/L, 0.006 $\pm$ 0.012 g/L, and 0.001 $\pm$ 0.001 g/L for TP, ALB, GLU, UA, respectively, which is acceptable for plasma analysis.

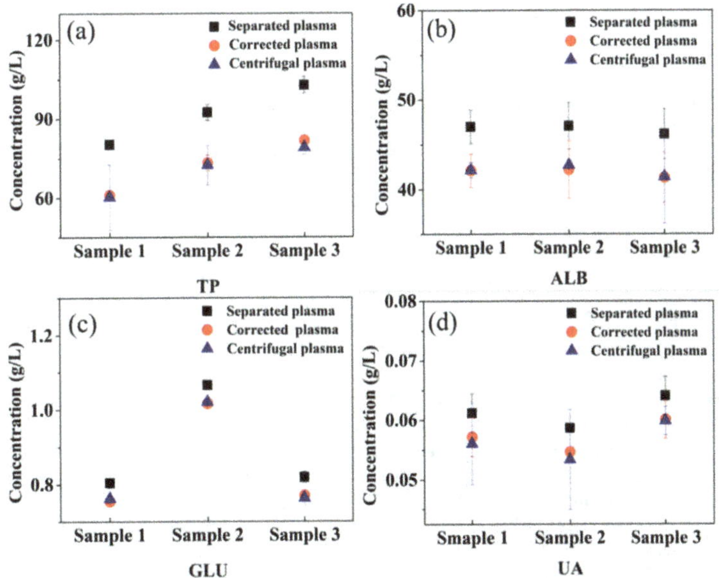

Figure 5. Component concentrations of clinical samples: (**a**) TP, (**b**) ALB, (**c**) GLU, and (**d**) UA.

4. Conclusions

We present a simple plasma separation device containing three layers of microspheres with different sizes as the separation barrier. It takes about 55 min to extract 3 µL of plasma from a 50 µL blood sample with this device. The evaporation and the absorption of water by the protein blocking coating of the microspheres during the separation process are the main causes of increased component concentrations in the plasma. Correction factors are applied to the device to eliminate those two factors. The feasibility of this device for clinical biochemical testing applications is validated with clinical blood samples for the measurement of TP, ALB, GLU, and UA in the separated plasma samples. The concentration differences between the separated plasma with our device and the centrifugal plasma after correction are 1.408 ± 1.056 g/L for TP, 0.168 ± 0.765 g/L for ALB, 0.006 ± 0.012 g/L for GLU, and 0.001 ± 0.001 g/L for UA, respectively, which are acceptable for POC analysis. This cost-effective, portable and external force-free plasma separation microchip shows great potential in POC analysis of blood biochemistry.

Supplementary Materials: The following are available online at https://www.mdpi.com/article/10.3390/mi12050487/s1, Video S1: the separated plasma passes through the capillary channels.

Author Contributions: Conceptualization, J.D., N.H., J.Q., L.G. and J.Y.; methodology, H.X. and Z.W.; validation, H.X. and Z.W.; formal analysis, H.X. and Z.W.; writing—original draft preparation, H.X.; writing—review and editing, H.X., J.D. and J.Y.; visualization, H.X.; supervision, J.Y.; project administration, J.Y.; funding acquisition, J.D., N.H. and J.Y. All authors have read and agreed to the published version of the manuscript.

Funding: This work was funded by the National Natural Science Foundation of China (Nos. 21827812, 81871450), the Natural Science Foundation of Chongqing, China (Nos. cstc2020jcyj-msxmX0680, cstc2019jcyj-bshX0006), and the Fundamental Research Funds for the Central Universities (2020CDJQY-A058, 2020CDJ-LHZZ-031).

Institutional Review Board Statement: The study was conducted according to the guidelines of the Declaration of Helsinki, and approved by the Ethics Committee of protocol code CZLS2021100-A on April 22, 2021.

Informed Consent Statement: Informed consent was obtained from all subjects involved in the study.

Conflicts of Interest: The authors declare no conflict of interest.

References

1. Mielczarek, W.S.; Obaje, E.A.; Bachmann, T.T.; Kersaudy-Kerhoas, M. Microfluidic blood plasma separation for medical diagnostics: Is it worth it? *Lab Chip* **2016**, *16*, 3441–3448. [CrossRef]
2. Yang, X.; Forouzan, O.; Brown, T.P.; Shevkoplyas, S.S. Integrated separation of blood plasma from whole blood for microfluidic paper-based analytical devices. *Lab Chip* **2012**, *12*, 274–280. [CrossRef]
3. Kersaudy-Kerhoas, M.; Sollier, E. Micro-scale blood plasma separation: From acoustophoresis to egg-beaters. *Lab Chip* **2013**, *13*, 3323–3346. [CrossRef]
4. Ammerlaan, W.; Trezzi, J.P.; Lescuyer, P.; Mathay, C.; Hiller, K.; Betsou, F. Method validation for preparing serum and plasma samples from human blood for downstream proteomic, metabolomic, and circulating nucleic acid-based applications. *Biopreserv. Biobank* **2014**, *12*, 269–280. [CrossRef]
5. Jung, W.E.; Han, J.; Choi, J.-W.; Ahn, C.H. Point-of-care testing (POCT) diagnostic systems using microfluidic lab-on-a-chip technologies. *Microelectron. Eng.* **2015**, *132*, 46–57. [CrossRef]
6. Nasseri, B.; Soleimani, N.; Rabiee, N.; Kalbasi, A.; Karimi, M.; Hamblin, M.R. Point-of-care microfluidic devices for pathogen detection. *Biosens. Bioelectron.* **2018**, *117*, 112–128. [CrossRef] [PubMed]
7. Tripathi, S.; Varun Kumar, Y.V.B.; Prabhakar, A.; Joshi, S.S.; Agrawal, A. Passive blood plasma separation at the microscale: A review of design principles and microdevices. *J. Micromechanics Microengineering* **2015**, *25*. [CrossRef]
8. Xia, Y.; Si, J.; Li, Z. Fabrication techniques for microfluidic paper-based analytical devices and their applications for biological testing: A review. *Biosens. Bioelectron.* **2016**, *77*, 774–789. [CrossRef]
9. Songjaroen, T.; Dungchai, W.; Chailapakul, O.; Henry, C.S.; Laiwattanapaisal, W. Blood separation on microfluidic paper-based analytical devices. *Lab Chip* **2012**, *12*, 3392–3398. [CrossRef]
10. Li, H.; Han, D.; Pauletti, G.M.; Hegener, M.A.; Steckl, A.J. Correcting the effect of hematocrit in whole blood coagulation analysis on paper-based lateral flow device. *Anal. Methods* **2018**, *10*, 2869–2874. [CrossRef]
11. Sista, R.S.; Ng, R.; Nuffer, M.; Basmajian, M.; Coyne, J.; Elderbroom, J.; Hull, D.; Kay, K.; Krishnamurthy, M.; Roberts, C.; et al. Digital Microfluidic Platform to Maximize Diagnostic Tests with Low Sample Volumes from Newborns and Pediatric Patients. *Diagnostics* **2020**, *10*, 21. [CrossRef] [PubMed]

12. Chen, X.; Cui, D.F.; Liu, C.C.; Li, H. Microfluidic chip for blood cell separation and collection based on crossflow filtration. *Sens. Actuators B Chem.* **2008**, *130*, 216–221. [CrossRef]
13. Crowley, T.A.; Pizziconi, V. Isolation of plasma from whole blood using planar microfilters for lab-on-a-chip applications. *Lab Chip* **2005**, *5*, 922–929. [CrossRef]
14. Han, J.Y.; DeVoe, D.L. Plasma isolation in a syringe by conformal integration of inertial microfluidics. *Ann. Biomed. Eng.* **2021**, *49*, 139–148. [CrossRef] [PubMed]
15. Robinson, M.; Marks, H.; Hinsdale, T.; Maitland, K.; Cote, G. Rapid isolation of blood plasma using a cascaded inertial microfluidic device. *Biomicrofluidics* **2017**, *11*. [CrossRef]
16. Rafeie, M.; Zhang, J.; Asadnia, M.; Li, W.; Warkiani, M.E. Multiplexing slanted spiral microchannels for ultra-fast blood plasma separation. *Lab Chip* **2016**, *16*, 2791–2802. [CrossRef] [PubMed]
17. Tsou, P.-H.; Chiang, P.-H.; Lin, Z.-T.; Yang, H.-C.; Song, H.-L.; Li, B.-R. Rapid purification of lung cancer cells in pleural effusion through spiral microfluidic channels for diagnosis improvement. *Lab Chip* **2020**, *20*, 4007–4015. [CrossRef]
18. Yang, S.; Undar, A.; Zahn, J.D. A microfluidic device for continuous, real time blood plasma separation. *Lab Chip* **2006**, *6*, 871–880. [CrossRef]
19. Maria, M.S.; Kumar, B.S.; Chandra, T.S.; Sen, A.K. Development of a microfluidic device for cell concentration and blood cell-plasma separation. *Biomed. Microdevices* **2015**, *17*. [CrossRef]
20. Shim, J.S.; Browne, A.W.; Ahn, C.H. An on-chip whole blood/plasma separator with bead-packed microchannel on COC polymer. *Biomed. Microdevices* **2010**, *12*, 949–957. [CrossRef]
21. Shim, J.S.; Ahn, C.H. An on-chip whole blood/plasma separator using hetero-packed beads at the inlet of a microchannel. *Lab Chip* **2012**, *12*, 863–866. [CrossRef] [PubMed]
22. Li, C.; Liu, C.; Xu, Z.; Li, J. A power-free deposited microbead plug-based microfluidc chip for whole-blood immunoassay. *Microfluid. Nanofluidics* **2012**, *12*, 829–834. [CrossRef]
23. Chen, M.-D.; Yang, Y.-T.; Deng, Z.-Y.; Xu, H.-Y.; Deng, J.-N.; Yang, Z.; Yang, J. Microchannel with stacked microbeads for separation of plasma from whole blood. *Chin. J. Anal. Chem.* **2019**, *47*, 661–668. [CrossRef]
24. Guo, W.; Hansson, J.; van der Wijngaart, W. Synthetic Paper Separates Plasma from Whole Blood with Low Protein Loss. *Anal. Chem.* **2020**, *92*, 6194–6199. [CrossRef] [PubMed]
25. Dixon, C.; Lamanna, J.; Wheeler, A.R. Direct loading of blood for plasma separation and diagnostic assays on a digital microfluidic device. *Lab Chip* **2020**, *20*, 1845–1855. [CrossRef]
26. Dimov, I.K.; Basabe-Desmonts, L.; Garcia-Cordero, J.L.; Ross, B.M.; Park, Y.; Ricco, A.J.; Lee, L.P. Stand-alone self-powered integrated microfluidic blood analysis system (SIMBAS). *Lab Chip* **2011**, *11*, 845–850. [CrossRef]
27. Kim, J.H.; Woenker, T.; Adamec, J.; Regnier, F.E. Simple, miniaturized blood plasma extraction method. *Anal. Chem.* **2013**, *85*, 11501–11508. [CrossRef]
28. Baskurt, O.K.; Meiselman, H.J. Blood rheology and hemodynamics. *Semin. Thromb. Hemost.* **2003**, *29*, 435–450. [CrossRef]
29. Kumar, R.; Nandhini, L.P.; Kamalanathan, S.; Sahoo, J.; Vivekanadan, M. Evidence for current diagnostic criteria of diabetes mellitus. *World J. Diabetes* **2016**, *7*, 396–405. [CrossRef]
30. Association Diabetes American. Updates to the standards of medical care in diabetes-2018. *Diabetes Care* **2018**, *41*, 2045–2047. [CrossRef]

Article

Innovative 3D Microfluidic Tools for On-Chip Fluids and Particles Manipulation: From Design to Experimental Validation

Sofia Zoupanou [1,2], **Maria Serena Chiriacò** [1,*], **Iolena Tarantini** [2] **and Francesco Ferrara** [1,3,*]

1   CNR NANOTEC—Institute of Nanotechnology, via per Monteroni, 73100 Lecce, Italy; sophia.zoupanou@nanotec.cnr.it
2   Department of Mathematics & Physics E. de Giorgi, via Arnesano, University of Salento, 73100 Lecce, Italy; iolena.tarantini@unisalento.it
3   STMicroelectronics S.R.L., via per Monteroni, 73100 Lecce, Italy
*   Correspondence: mariaserena.chiriaco@nanotec.cnr.it (M.S.C.); francesco.ferrara@st.com (F.F.)

Abstract: Micromixers are essential components in lab-on-a-chip devices, of which the low efficiency can limit many bio-application studies. Effective mixing with automation capabilities is still a crucial requirement. In this paper, we present a method to fabricate a three-dimensional (3D) poly(methyl methacrylate) (PMMA) fluidic mixer by combining computer-aided design (CAD), micromilling technology, and experimental application via manipulating fluids and nanoparticles. The entire platform consists of three microfabricated layers with a bottom reservoir-shaped microchannel, a central serpentine channel, and a through-hole for interconnection and an upper layer containing inlets and outlet. The sealing process of the three layers and the high-precision and customizable methods used for fabrication ensure the realization of the monolithic 3D architecture. This provides buried running channels able to perform passive chaotic mixing and dilution functions, thanks to a portion of the pathway in common between the reservoir and serpentine layers. The possibility to plug-and-play micropumping systems allows us to easily demonstrate the feasibility and working features of our device for tracking the mixing and dilution performances of the micromixer by using colored fluids and fluorescent nanoparticles as the proof of concept. Exploiting the good transparency of the PMMA, spatial liquid composition and better control over reaction variables are possible, and the real-time monitoring of experiments under a fluorescence microscope is also allowed. The tools shown in this paper are easily integrable in more complex lab-on-chip platforms.

Keywords: 3D microfluidics; lab-on-a-chip; micromixer; gradient generation; polymeric device; buried channels; particles manipulation tool

Citation: Zoupanou, S.; Chiriacò, M.S.; Tarantini, I.; Ferrara, F. Innovative 3D Microfluidic Tools for On-Chip Fluids and Particles Manipulation: From Design to Experimental Validation. *Micromachines* **2021**, 12, 104. https://doi.org/10.3390/mi12020104

Received: 21 December 2020
Accepted: 19 January 2021
Published: 21 January 2021

1. Introduction

During the last decades, microfluidic structures have represented the cornerstone of common lab-on-a-chip (LOC) devices. Conventional fabrication methods include the use of standard soft lithography, followed by PDMS/PDMS, PDMS/glass, and PDMS/SU8 bonding protocols to obtain assembled microfluidic devices [1–4]. Despite the good performance of all the aforementioned molded devices, their suitability for biological studies, and the ease of process, they all are facing some limitations, which makes them ineligible for further repeated analysis and assessment, since they do not meet the requirements of robustness and stability and do not allow (in many cases) a strong reproducibility of biological or chemical assays [5,6]. Moreover, materials like polymeric silicones and resins listed above are not suitable for industrial exploitation and inappropriate out of a research laboratory [7]. Such limitations include bond strength issues, deformability, loss of hydrophilicity, lack of long-term sealing stability, and strong evaporation phenomena [8–11]. In addition, PDMS is sensitive to exposure to some chemicals and may adsorb proteins on its surface [12–14]. Moreover, in the realization of more complex and three-dimensional (3D) microfluidic devices, the apposition of different layers amplifies these drawbacks, as it is difficult to maintain the alignment and the good sealing of microchannels without deforming them and then affecting microfluidic features.

Thermoplastic polymers (plastics) like poly(methyl methacrylate) (PMMA), polystyrene (PS), cyclic olefin copolymers (COC), or polycarbonate (PC) have gained increasing interest during the last decade, as they allow easy surface treatment/modification, are generally transparent and biocompatible [5] and demonstrate suitability to meet some of the industrial requirements for the LOC market [15]. In particular, thanks to their low costs, they can be used for the fabrication of disposable microfluidic chips, opening a way to a wide plethora of applications, ranging from mixing [16], sample preparation, and analytical tools [17].

On the other hand, the possibility to switch from planar to 3D microfluidics has emerged as a potential revolutionary technology due to the unique properties of these miniaturized fluidic systems for their use in cell biology and on-chip chemistry to allow innovative methods like droplet microfluidics [18] and on-chip cytofluorimetry [19]. The innovation of 3D designs has gained significant attention and has been established as an important tool for different laboratory applications [20–22], if the entire structure is realized in transparent materials [23] and allows the real-time observation of phenomena in different and separate microchannels at the same time in a single microscopic frame.

A particularly crucial element to shift standard assays to on-chip reactions are micromixers which have a wide range of bio-applications such as studies on living cells for medical diagnostics, nanoparticles synthesis for chemistry, and biotechnologies analysis like polymerase chain reaction (PCR) [24–26]. In this respect, many research groups have focused their attention on this aim. Rasouli et al., as an example, designed and studied a curved micromixer channel with obstacles, in order to create normal advection and to generate Dean vortices through standard photolithography [27]. In that case, the patterned PDMS was bonded with glass. Baccouche et al. realized another microfluidic mixing device which is applied to the creation of droplets by mixing elements of two to four different tubes [28]. In another very recent study, the authors described the design of a fluidic mixer which utilizes air bubbles to facilitate mixing [4].

To combine the needs of robustness and reproducibility with the possibility to check a complex experiment into a single device making use of 3D microfluidics, we realized a stable, transparent and plug-and-play device made of PMMA, developed into three different levels. The 3D micromilled microfluidic structure is based on a PMMA/PMMA bond realized by a low-cost and easy process recently described to obtain a circulating tumor cell (CTC) device in a planar system [17]. The resulting experimental system consists of three PMMA substrates, all obtained by a micromilling machine with the following components: (i) a downstream reservoir/mixing channel on the bottom to increase the efficiency of diffusion, (ii) an upstream curved channel section, and (iii) a layer with an inlet and an outlet on the top, holding holes for firm microtubes connection and not being affected by pressure and flow strains, without the need for additional glues or gaskets for watertight connections. The pivotal element of our device is the buried transition region which creates a connection point between two of the three vertical layers by a hole between two microchannels. Since our strategy is based on sealing patterned PMMA substrates, a critical further step during the fabrication process was the bonding of the microstructured substrates. The method applied is an innovative thermal and solvent-assisted bonding, which makes use of low temperatures, low pressures, and a very common solvent (isopropanol) to allow us to obtain strong adhesion among layers, resulting in a monolithic structure with no wake points related to the typical sealing problems of microfluidic devices. Indeed, glues, clamps, or additional luer fittings and gaskets are not necessary for a watertight seal. Moreover, no burrs, cracks, or recast layers occur during the bonding process, so neither the sealing procedure nor the transparency is affected for microscope observation.

The realization of this tool is described in next paragraphs from CAD design, optimization by the finite element method (FEM) simulation, prototyping, and final experimental tests obtained by the injection of colored liquids and differently labelled nanoparticles. 3D microchannels were used in two-way flow directions to use the two shaped microchannels (serpentine and reservoir), and this allowed us to prove the efficiency of the device for further applications for gradient creation and mixing performance mimicking macroscale turbulence [29,30]. The chip has the potential to be a tool for mixing with application in chemistry and cell biology. In cell biology studies, such a device allows for the activation of normal fibroblasts by external stimuli to become cancer-associated fibroblasts (CAFs) [31], the mixing of CAFs and associated tumor cells for interplay investigations, and the immunomagnetic labelling of circulating tumor cells to enrich a diluted sample [32]. In biochemistry, the chip could be a valid alternative to prepare a low volume of particular drugs which need "on demand" or in-situ preparation (toxic or costly reagents) to work with droplets [33], to perform digital PCR or to study protein–nanoparticles interactions [34].

2. Materials and Methods

2.1. Materials

Fluidics and particle tracing of interconnected h-junction microchannels in three dimensions were simulated with Comsol Multiphysics 5.1 computational package (COMSOL, Inc., Burlington, MA, USA), using a Computational Fluid Dynamics CFD module and specifically "mixture model, laminar flow". During the entire simulation, the particles' diameters were defined at 200 nm and 1 μm.

For the fabrication of the device, we used 3 square PMMA substrates (Vistacryl CQ; Vista Optics, Gorsey Lane, Widnes, Cheshire, WA8 0RP, UK), with both the length and the width of 30 mm and a thickness of 1 mm for the substrates, which hosted micromilled channels, and a thickness of 2.5 mm for the substrate, which served for the holes.

For microchannels bonding, we used pure isopropyl alcohol (Sigma-Aldrich, St. Louis, MO, USA), and for microchannels functionalization, we followed a sequence of steps which included O_2 plasma surface treatment and the incubation of the device with 1 mg/mL bovine serum albumin (BSA) (1%) in Phosphate Buffered Saline (PBS) buffer) (Sigma-Aldrich, St. Louis, MO, USA).

The Elveflow microfluidic setup (Elvesys, Paris, France), which is suitable for these kinds of studies, was installed for pumping the solution into the device. For image and film acquisition, we used an Axio Zoom V16 fluorescence microscope (Zeiss, Oberkochen Germany) with an Apo Z 1× objective. The eyepiece of the microscope was 10×/23Br, and the numerical aperture (NA) was 0.25. The two validation tests were performed with fluorescent polystyrene microspheres of 200 nm (green) and 1 μm (red) in size (FluoSpheres® Fluorescent Microspheres, Invitrogen, Ltd. 3 Fountain Drive Inchinnan Business Park, Paisley PA4 9RF, UK) for particles mixing, and ethanol (Sigma-Aldrich, St. Louis, MO, USA) and colored ink for colored liquids mixing were used. The wavelengths of fluorescence microscope channels were set according to the nanoparticles' provider, i.e., for green particles, the excitation and emission maxima were 505 and 515 nm, respectively), and for red particles, the excitation and emission maxima were 660 and 680 nm, respectively.

2.2. Computational Modeling, Fabrication, Functionalization, and Testing of the Chip

2.2.1. Modeling

The first step, in the scope of this work, was to model the device. To this end, we designed a 3D network, consisting of two interconnected microchannels having two independent inlets and sharing a common outlet. The COMSOL Multiphysics was then used to simulate the fluidics and particle tracing and to evaluate the mixing response of the interconnected microchannels. For the simulation, a free tetrahedral mesh was used for the entire microchannels. For the boundary conditions, we selected identical inlet velocities/flow rates, and zero static pressure was applied at the outlet. At the channels' walls, the applied flow condition was zero, and inside the network of channels, we added water in all domains for species transport. Furthermore, we required the flow to be incompressible, steady and with no gravitational effects at all domains of the design. As a next step, we set up all the necessary conditions for particles mixing by selecting two different inlets and a common outlet and applying particle parameters from standard polystyrene beads. We also took into consideration the drag and gravitational forces. These conditions were kept constant during the whole simulation. After setting the aforementioned parameters, we evaluated the performance of the design by firstly checking the distribution of the flow at the domains of the design, but also the velocity, the flow rate, and the pressure at the inlets, junction point, and the outlet. Finally, for particles simulations, the results for both simulations, i.e., mixing two different kinds of particles and diluting a specific number of particles, were satisfactory. The numerical model used has been validated against a real system implementation.

2.2.2. Design and Fabrication

The bottom, middle and top layers with reservoir, serpentine and inlets/outlet holes respectively were designed using Solidworks CAD software (SolidWorks Corporation, 300 Baker Avenue, Concord, MA, USA), and computer-aided manufacturing (CAM) software was used to transfer the CAD information in a machine code file for the micromilling control. The Mini-Mill/GX micromilling machine (Minitech Machinery, Norcross, GA, USA) was used to realize channels and inlet/outlet holes in the PPMA layer by a 200 µm 2-flute carbide micro end milling tool. To obtain channels with a 400 µm width, a 200 µm height, and a good surface roughness, the feed rate was set at 150 mm/min with a spindle speed of 20,000 rpm. The layers were aligned using an on-board camera.

2.2.3. Sealing

In order to assemble the device and to achieve a robust bonding between the PMMA layers, it was necessary to deposit hot, pure isopropyl alcohol (70 °C) on a flat bottom substrate for 10 s with a spin coater set at 2000 rpm. As soon as the bonding was established, we used binder clips to make firm alignment and incubated it for 20 min at 60 °C, without additional pressure in order to improve the bonding.

2.2.4. Functionalization

The PMMA substrates suffered from the high hydrophobicity of the surface. To mitigate this, right after assembling the PMMA slices, we treated them with O_2 plasma, which is a known method for improving surface wettability and for increasing hydrophilicity, allowing for the easier flowing of the subsequent water-based solutions. All the steps of functionalization and sample injection were performed directly in-flow, thanks to the perfect fitting of the holes with capillary tubes, allowing for a plug-and-play way of using the device. Tubes, indeed, were firmly connected, without the need for additional glues, gaskets or clamps to ensure the watertight sealing of connections. The solutions were then flowed into the capillary tubes and channels. To attenuate any unspecific sticking of fluorescent polystyrene particles on the PMMA channels surface, we incubated the chip with a blocking buffer (1 mg/mL BSA in PBS) for 2 h at room temperature.

2.2.5. Experimental Tests

To evaluate the microfluidic device, we implemented two kinds of experiments for mixing and gradient generation with both colored fluids and particles. In addition, in some cases, we used on purpose the outlet as an inlet to test the influence of different path lengths and shapes of the microchannels on the mixing capabilities of the device.

During the first series of the experiments, we used two different colored liquids, which were loaded into the capillary tubes to be injected on the top (serpentine) and the bottom (reservoir) channels simultaneously. For the second series of the experiments, we injected two kinds of particles, that is, one had a diameter of 200 nm and the other had a diameter of 1 µm. This allowed us to observe the mixing behavior of liquids and particles. Both experiments were performed at different flow rates. To deliver the sample into the device, we connected the device to the Elveflow microfluidic setup by utilizing perfectly fitting capillary tubes with micromilled shaped inlets and outlets in both cases. The Elveflow micropumping system was equipped with an OB1 base module, two MkIII+ channels for a pressure controller, and two microfluidic flow sensors. With this microfluidic system, the flow of the medium can be controlled temporally. During the experiments, the two inlets were connected with a vial containing the particles samples. Subsequently, the microchip was positioned under the microscope for monitoring, and the result was validated with real-time image acquisition.

3. Results

3.1. Verification of the Numerical Model

The conceptual design of the 3D microfluidic tool has been the first step of our work. The innovative features that were required for our device led us to investigate the fluidic parameters at stake before planning the final architecture. To this end, finite element simulations were conducted to predict the performance of a 3D fluidic micromixer under two different regimes, i.e., mixing and dilution of particles. The design of the interconnected channels and the simulation results for the mixing index are given in Figure 1. In order to have a better insight on the performance of the mixing design, the flow behavior was investigated. Figure 1a reports the velocity distribution of the proposed design, showing a maximal flow at the point where the two channels meet and a minimal flow originated from the inlets till the junction point. At the channels side walls, the velocity was infinitesimal. Furthermore, Figure 1b shows the levels of the pressure at the entire network. In our design, the pressure strength was reduced from the channels connection point toward the common outlet, which was expected due to the inverse proportion to the velocity. The flow rate, which is relative to the velocity, was calculated through the given equation:

$$Q = A\bar{u} \tag{1}$$

where Q is the flow rate, A indicates the cross-sectional area, and $\bar{u}$ is the average velocity.

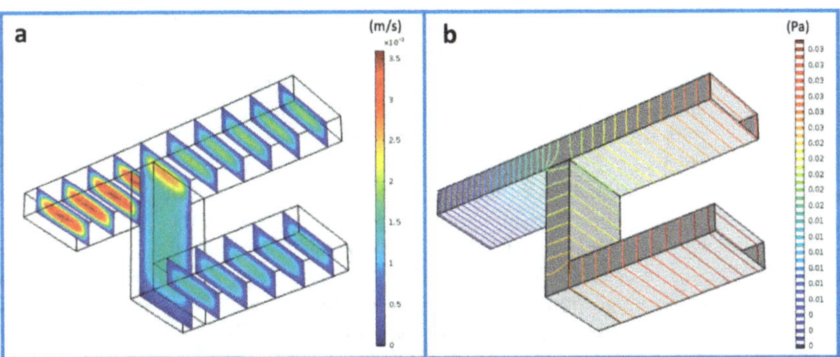

Figure 1. Finite element method (FEM) simulation of flow velocity (**a**) and pressure (**b**) in a two-level interconnected microchannel system.

Hence, for each experiment, i.e., mixing and diluting, we performed a sequence of simulations where we tweaked the values of some parameters in between. These parameters included the flow rate and the pressure. For the mixing experiment, a suspension of particles with sizes of 1 μm and 200 nm was injected in both inlets, respectively, and the values of flow rate ranged from 1 mL/min up to 5 mL/min between each iteration (Figure 2a–d). The resulting product was a mix of the two populations as expected in the range time of 0–67 s.

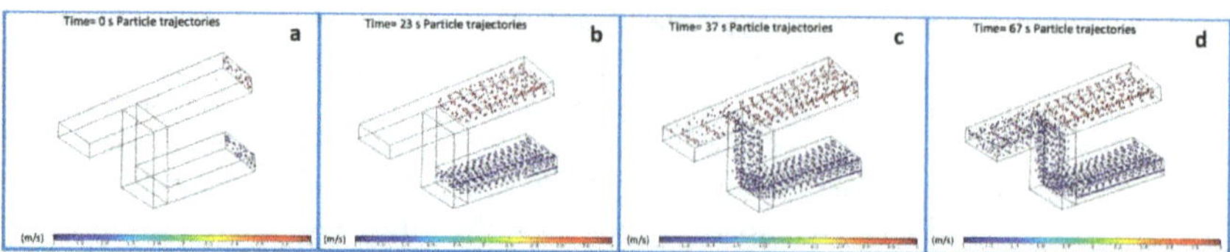

Figure 2. Frames of the simulation of particles mixing into three-dimensional (3D) microchannels from the starting point at t = 0 s (**a**), 23 s (**b**), 37 s (**c**) to 67 s (**d**) when particles are completely mixed.

To simulate the dilution experiment, we replaced the solution in one of the inlets with water. In Figure 3, it is reported that the particles approaching the outlet were indeed compelled to be diluted due to the presence of water on the bottom channel. Videos showing simulated behaviors for both particles mixing (Video S1) and dilution (Video S2) are provided in Supplementary Materials.

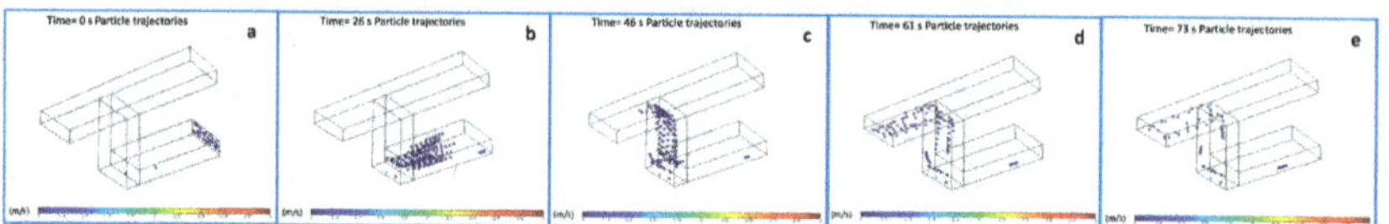

Figure 3. Frames of the simulation of particles dilution from the starting point at t = 0 s (**a**), 26 s (**b**), 46 s (**c**), 61 s (**d**) to 73 s (**e**).

3.2. Design and Fabrication of the LOC Device

In order to verify the accuracy of our numerical model in predicting the mixing process, a real mixing chip was fabricated. Selecting the proper design parameters that have the greatest influence on the mixing quality is a crucial issue for a chip made on purpose. As a first step in the realization of the device, we designed the CAD file of its elements, which was then transferred in machine code to the fabrication instruments used. Figure 4 schematically illustrates the proposed geometry. In particular, with the aim to have a structure which could be useful for both kinds of experiments, namely mixing and dilution proofs of concepts, we decided to consider a reservoir chamber in the bottom channel and a serpentine-shaped top pathway. Therefore, the entire device was constituted by a three-level structure: a reservoir-shaped channel in the first, a serpentine-shaped channel in the central, and three holes as inlets and outlet in the top layer. The three layers were separately micromilled using a mechanical micromilling machine mounting a 200 μm tool, by carefully considering alignment markers on the three layers. The length of the serpentine was of around 7 cm organized in 6 loops; the total length of the bottom channel was of around 5 cm, and the diameters of the oval reservoir were of 2 and 4 mm; the common portion ran for 1.5 cm. The substrate containing the serpentine-shaped channel held a through-hole to obtain the buried junction between the channels and to allow a 3D microfluidic pathway. The holes in the upper layer were shaped to host the capillary tubes to allow for a plug-and-play connection ensuring tight connections.

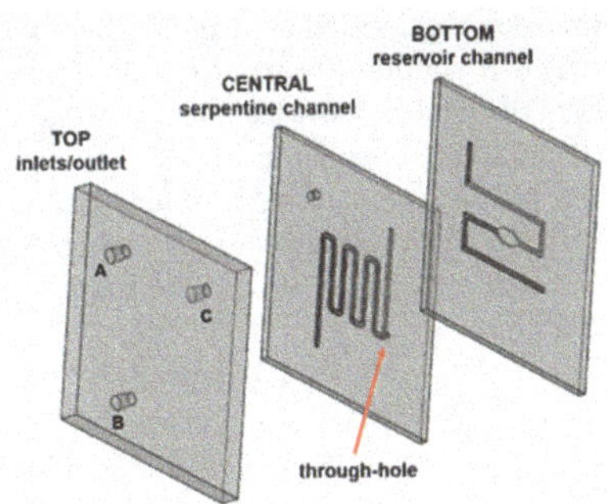

Figure 4. Exploded structure of the 3D microfluidic tool for the mixing and dilution of particles. The three layers (bottom, central, and top) hold the reservoir-shaped channel, the serpentine-shaped channel with a through-hole, and the inlets and outlet (indicated by A, B, and C), respectively.

3.3. Bonding and Functionalization of the Device

Once the PMMA layers were fabricated, as a first step, we bonded together the substrates with microchannels; then on the top of them, we placed the slice with the holes, following the alignment scheme described above. The sealing approach used in this work was a thermal- and solvent-assisted bonding method, where hot isopropyl alcohol was spin-coated on the surfaces of the first two substrates. Then, the wet substrates were quickly aligned and firmly pressed with binder clamps. Finally, shortly after, the assembled PMMA slices were put in an oven, providing at the end of the procedure an irreversible bonding, which was solely based on this easy technique [17]. The same procedure was then reiterated to align the substrate with the holes to the former two. At the end of the procedure, the device resulted in a stable monolithic chip, with buried channels and an interconnection hole and with the inlets and outlet on the upper layer.

After assembly, the device underwent the O_2 plasma treatment to improve hydrophilicity of the PMMA channels. Finally, we connected the device to the Elvesys micropumping system with the capillary tubes to test the bonding quality and the leakage of the chip by systematically increasing the pressure from 10 to 800 mbar while performing the monitoring via an optical microscope. We performed the in-flow functionalization with BSA to avoid polystyrene particles to stick on the channels wall and to maintain the hydrophilicity through time [35,36].

3.4. Mixing and Gradient Generation Experiments

With highly efficient mixing of different elements being the ultimate goal, the micromixer was tested by performing a flow test. In Table 1, a summary of the experiments we performed is reported, and the holes (A, B, and C) were used as inlets or outlets.

Table 1. Scheme of experiments based on the alternative use of holes as inlets/outlets combined with the injection of colored fluids or fluorescent particles suspensions.

Holes	Mixing Experiment			Gradient Generation	
	Colored Fluids (#1)	Colored Fluids (#3)	Particle Suspension (#4)	Colored Fluids (#2)	Particle Suspension (#5)
A	in	In	in	in	out
B	in	Out	in	in	In
C	out	in	out	out	in

First and foremost, we decided to visualize the mixing behavior by using two different colored liquids. This first attempt (experiment #1 according to the scheme in Table) can reflect the fitness of the device. Flow tests were carried out using the Elveflow microfluidic setup. Once the capillary tubes were filled with the colored liquids, we injected them through the inlets, following the filling of microchannels by an Axiozoom microscope. In order to make the two fluids contemporarily reach the common portion of the channel and to obtain a chaotic particle motion at the mixing point [29], we initially set the flow parameters at 3 μL/min.

Based on different resistances and velocities of the flow (due to the different shape of the microchannels and different numbers of 90° bends of the two paths), we carefully tuned the parameters during the experiments. The long way the fluid went along, together with the large field of view of the microscope (around 10 mm) allowed us to keep eyes on the channel filling, and we reached the optimized control and mixing capabilities at a flow rate of 2.08 μL/min and a pressure of 42.80 mbar for channel 1 (serpentine) and at a flow rate of 1.65 μL/min and a pressure of 63.08 mbar for channel 2 (reservoir). The Figure 5a illustrates an overall picture of the microfluidic chip during the in-flow test we executed, in which we can observe also by naked eye that the injected pink and blue fluids resulted in a violet mixed color at the outlet. Apart from that, a more clear image obtained by the microscope is shown in Figure 5b.

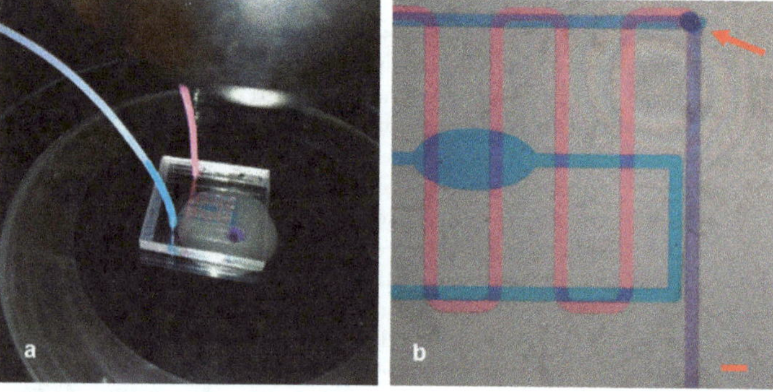

Figure 5. (**a**) Picture of the experimental setup for experiment #1. The blue drop emerging from the outlet of the assembled and connected device is clearly visible as a result of the mixed pink and blue fluids. (**b**) Image of the separate and common pathway taken through the microscope. Scale bar: 1 mm.

The pink solution was infused in the upper serpentine channel, and the reservoir was filled with the blue one. In Figure 5b, it was vividly observable that each liquid followed its separate path and that eventually they merged at the meeting point as expected. The buried hole (indicated by the red arrow) was where the two fluids intersected and the common path began, corresponding to the very last portion of the serpentine. After that point, the two colors of the fluids were not distinguishable anymore, as they merged and turned violet.

Moreover, to test the capability of the device in a laminar fluid regime, we tuned the flow rate of the channels by enhancing in turn the two channels and making the pink or blue ink prevail on the other (experiment #2, Figure 6), resulting in a gradient generator tool.

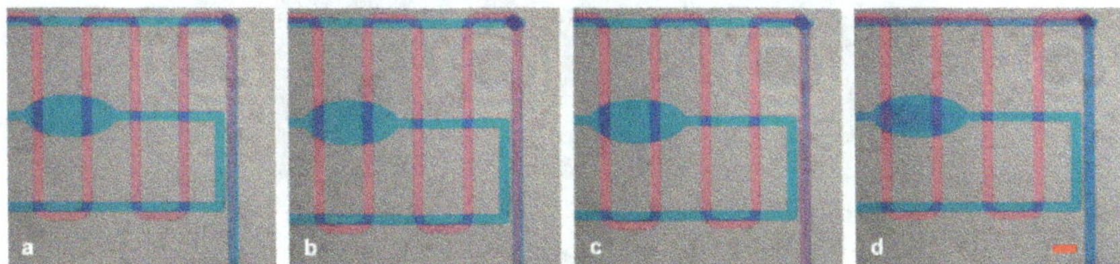

Figure 6. (a–d) Sequence of the pink or blue prevalence if the laminar flow is produced instead of mixing in experiment #2. Scale bar: 1 mm.

To demonstrate the mixing capabilities of the apparatus in a longer pathway, we switched positions between one inlet and the outlet, making the serpentine channel work as a mixing channel (Experiment #3, holes A and C as inlets). In this case, we filled the serpentine channel with pink fluid injected from hole C, and we made the blue fluid arrive from the reservoir channel (hole A) to the interconnection hole, in order to reach the serpentine channel. The results are demonstrated in Figure 7, in which the pink solution in the serpentine channel started to fade after being mixed with the blue liquid and showed a transient color switch to violet. The same procedure can be applied utilizing the reservoir channel as a mixing channel (holes B and C as inlets).

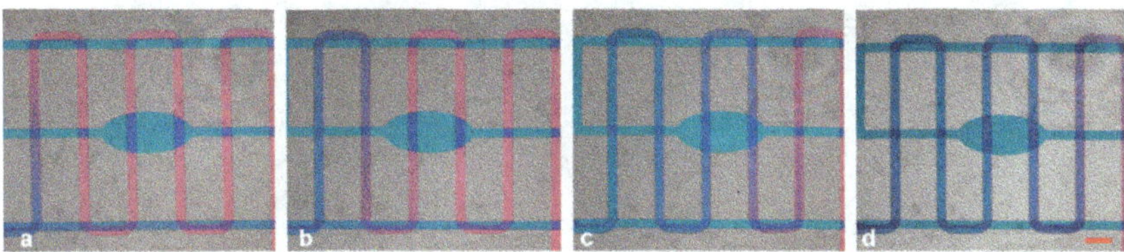

Figure 7. (a–d) Mixing capabilities of the serpentine-shaped channel introducing pink ink from the hole previously selected as an outlet in experiment #3. Scale bar: 1 mm.

After having tested the device for watertightness and mixing capabilities, we proceeded to the second phase of the experiments with the particles mixing experiment (experiment #4). We injected a suspension of 9.1×10^5/mL green fluorescent polystyrene particles of 200 nm in size to the reservoir channel through hole A and 7.2×10^5/mL of 1 µm particles to the serpentine channel through hole B and recorded the normal flow with a fluorescent microscope. As can be seen, while the two flows were activated independently, the particles were clearly separated and they can be observed contemporarily in a single microscope frame. Green fluorescence can be detected only in the bottom channel, while red fluorescence was observed only in the upper one without moving the device (Figure 8).

It is worth noting that consecutive experiments were executed in a wide range of flow rates, in order to test the mixing performance. Since a mutually proportionate flow rate in the channels is important, each change of the flow in a channel was accompanied by a corresponding pressure compensation of the other channel, due to the higher resistance at the serpentine path. The injections of the particles were performed with a flow rate and a pressure of 2.13 µL/min and 26.42 mbar, respectively, in channel 1 (serpentine, red particles) and with a flow rate and a pressure of 1.46 µL/min and 39.52 mbar, respectively, in channel 2 (reservoir, green particles) stabilized to contemporarily reach the common portion of the channel. The path of particles was followed in real time under the microscope, and some pictures were captured in consecutive moments immediately soon before and after the mixing point (Figures 8 and 9).

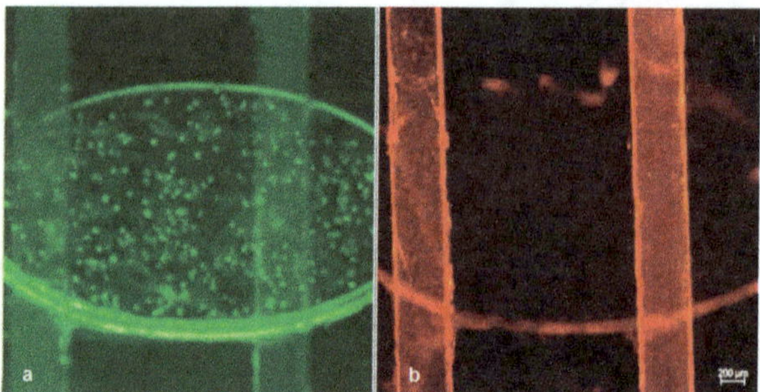

Figure 8. Experiment #4: merged image taken under a fluorescence microscope with subsequent fluorescence switching focused on the bottom reservoir channel (excitation/emission: 505/515 nm) filled with red nanoparticles (**a**) and the upper serpentine channel (excitation/emission: 660/680 nm) filled with red nanoparticles (**b**).

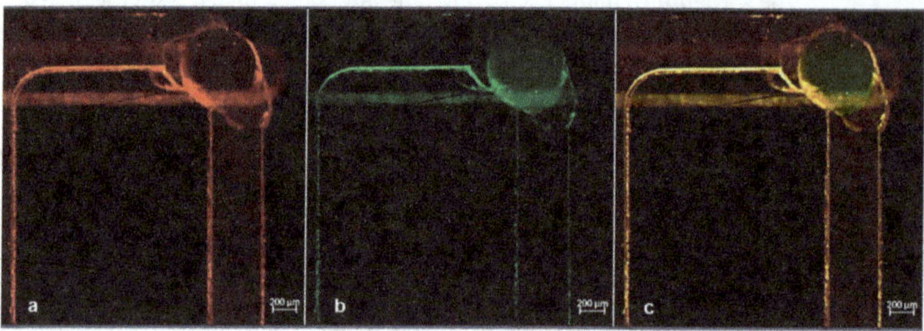

Figure 9. (**a,b**) Snapshot of the mixing point obtained by red and green fluorescence channels during experiment #4. The presence of red particles was clearly visible, while few particles were detected in the bottom and emerging flow. (**c**) Merged image of red and green channels.

The results of this mixing test are illustrated in Figure 9a–c, where the particles seemed to be mixed while moving from their individual channels to the common one. Figures 9a–c and 10a–c were acquired by selecting separately red and green fluorescence channels with the microscope and subsequently applying the "merge" command. In this way, we can very easily distinguish both streams of particles (green and red) arriving the common channel from their individual paths.

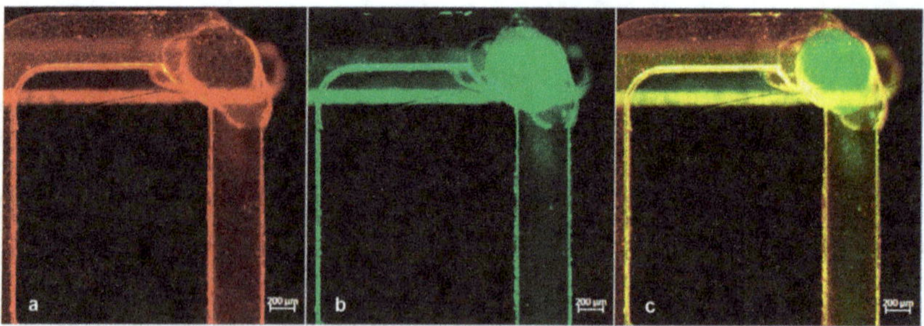

Figure 10. (**a,b**) Snapshot of the mixing point obtained by red and green fluorescence channels during experiment #4. The presence of red particles was clearly visible in the top serpentine layer, while in the bottom channel the amount of green particles was increased with respect to Figure 9b. (**c**) Merged image of the two fluorescence channels confirming the mixed particles solution in the common path.

Concretely, a comparison of Figure 9 with Figure 10 can aid to clarify that in the first, the particle mixing was just ignited since the presence of green particles was meager. In contrast, in Figure 10, we can see that a significant number of both particles were detected. This verified the mixing efficiency that needed to be met.

Further examination of the adoption of the device as a gradient generator tool was held using fluorescent particles (experiment #5). To this end, one channel was maintained with full of particles injected from the inlet of the reservoir channel (hole A); pure water was instead injected from hole C (previously considered as an outlet), running through the serpentine channel reversely. During this experiment, the same procedure was repeated to keep the water flow constant (around 7 µL/min with a pressure of 50 mbar) and gradually diminishing the particles flow between the iterations, in order to obtain a gradient generator. The values of the flow rates and the pressure for particles injection ranged from 7 µL/min (around 50 mbar) down to 1 µL/min (20 mbar) with steps of around 1.5 µL/min. Figure 11 depicts the progressive dilution of the particles. Specifically, in Figure 11a, the concentration of particles (initial concentration of 7.2×10^5/mL) was at the highest level, while we kept an equal flow rate for both solutions. Subsequently, a gradual decrease of particles concentration was observed, when we tuned the flow to reach the lowest-value set (Figure 11b,c). In this way, the collected fraction emerging from the outlet gradually diluted the suspension of particles from the first to the last. The final step tested was the complete stop of the particles channel (flow 0), which led to a fraction including the particles remaining in the channel and a progressive removal of unspecific adhered polystyrene beads.

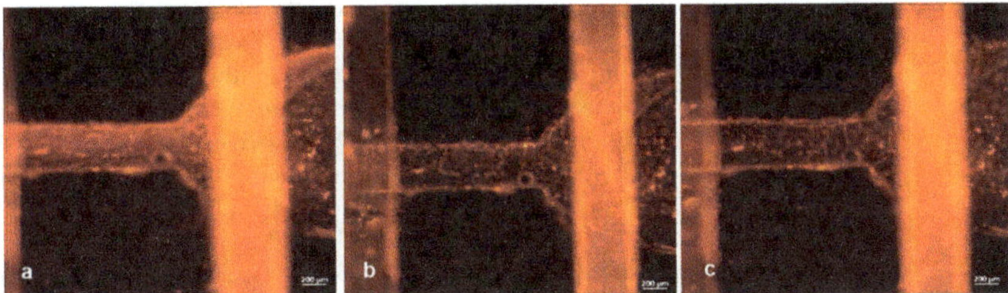

Figure 11. Experiment #5: on-chip particle dilution sequence. (**a**) The flow rates for particles and water injection were quite the same, and a high number of particles and associated fluorescence were clearly visible at the reservoir channel. (**b,c**) The flow rate of the particles was decreased, while keeping the water flow rate constant.

4. Discussion

Most major microfluidics applications demand quicker mixing, high accuracy, and low reagent volume. However, microfluidics and especially micromixing have not reached that level of efficiency yet. Recently, there has been an emergence of 3D microfluidic designs which promise to tackle these issues and streamline micromixing. Apart from that, 3D designs improve the mixing process and the possibility to perform high throughput tests: one device can be used for several parallel experiments. These are the reasons why we were encouraged to take advantage of this potential and designed and manufactured a PMMA microfluidic mixer.

In the present study, we firstly simulated the performance of an innovative 3D micromixer and then tested and ratified the modeled approach experimentally. This allowed us to identify weak points of the conceptual design and to understand how to modify the workflow in order to obtain the end-point results of our LOC tools.

In order to move into a real experimental setup, we used microfabrication by micromilling technology on PMMA substrates. Taking advantage of this highly flexible fabrication method, we were able to perfectly tune the design and to decide the best architecture for minimizing the limitations of the geometry, since it allows us to generate any feature or shape. In the right range of dimensions, micromilling has been proved to hold better performances in comparison with other fabrication methods. For example, photolithography, which is a multistep process, is almost impossible to execute without using a brand new master, which can limit the options at the prototyping phase. A device of a few square centimeters (3×3 cm^2) fabricated with high precision (the channel width of 400 μm in this case), short processing time (it takes few minutes for micromilling single layers) and relatively low costs is desirable, as PMMA substrates are among the cheapest materials for the purpose. Besides its great mechanical properties, transparency, low toxicity, biocompatibility, preserved stability throughout all the experiments, and the possibility to use a well-established technique for bonding multilayers PMMA gave us also the possibility to obtain a robust and reproducible monolithic device, featuring all the characteristics required to obtain 3D microfluidics. In particular, our device consisted of three overlapped and sealed layers. The strong point of the tool is that the lowest reservoir-shaped channel and the central serpentine ran separately for the most of their course but at a certain point they were interconnected by a through-hole, resulting in the realization of a common portion of the channel, which was able to host mixing reactions. To complete the architecture of the 3D device, the third top layer was assembled to ensure the inlets and outlet connections and micromilled to perfectly fit with the tubes without the need for any additional procedures to achieve watertight experiments.

Once assembled, the device was successfully used as a mixing and dilution tool for easy integration on LOC systems. Moreover, the perfect transparency of our tool allows the contemporary observation of separate events happening in the two channels, with a single microscope observation. In the proposed configuration, the two pathways exploit the same area to run, but they are separated for the most of their lengths. The design and geometry of the channels can be easily modified on the basis of the selected application. This enables the user to perfectly monitor the inner of the channels, without moving the device or loosing the point of view to check alternatively two (or more) different behaviors. To demonstrate this, we used two differently labelled particles suspensions, injected into the two buried microchannels, and we were able to follow the two populations by simply shifting the fluorescence channel of the microscope. Mixing and gradient generation experiments were performed firstly using colored fluids to check the flow capabilities and mixing properties of the buried interconnected channels, and after we tested the mashing of the fluorescently labelled polystyrene nanoparticles, a highly efficient mixing was obtained, as demonstrated by fluorescence analysis.

In the first case, we were able to finely tune the flow and velocity of the two channels in order to obtain a complete mixing of the two colored fluids or alternatively, to maintain a laminar flow. Moreover, reversing the flow by connecting the tube from the outlet hole, we were able to make the serpentine channel as a mixing channel, in order to exploit a longer path for mixing.

After establishing flow rates and pressure conditions for colored fluids, we started dealing with particles. We injected a suspension of 9.1×10^5/mL of green particles in the bottom reservoir-shaped channel and a suspension of 7.2×10^5/mL of red particles in the serpentine-shaped channel. We firstly tested their presence in channels running separately till the connection hole, where the particles were forced to mix running through the common portion of the 3D microchannels. Thanks to the complete transparency of the device, we were able to follow the entire experiment under a microscope and obtain good fluorescence pictures without any interference with the PMMA multilayers.

Finally, to prove the effectiveness of our device, we provided experimental conditions for diluting particles and creating a gradient of concentration and used the bottom reservoir channel as a mixing chamber for water and particles. We injected particles and water, maintained a constant water flow and decreased the flow rate of particles from 7 μL/min to 0 μL/min with steps of 1.5 μL/min. The collected fraction from the device's outlet was gradually more diluted over time. Based on the results of the experiments as proofs of concept, we showed that the apparatus can indeed host the expected functionality while maintaining high efficiency and accuracy.

This also proves that this LOC device's architecture could be considered a valuable tool for a plethora of applications that require high mixing efficiency and has the potential for integration into portable systems as well. Moreover, the high stability and robustness of the assembled device make it suitable for the on-demand mixing functions (mixing and labelling cells and nanoparticles, mixing drugs which are forced to be prepared "near-the-bed", etc.) and can enable standard microfluidic connections to be promptly merged for simple plug-and-play LOCs.

5. Conclusions

LOCs, defined as devices which include multiple laboratory features within a few square centimeters, have the breakthrough potential for radically changing traditional methods in various fields of chemistry, environment, and life sciences [37–39]. Rapid prototyping methods to design and realize polymeric LOCs are on the rise, as they allow high flexibility and precision. The fabrication of a microfluidic platform, for example, may require optimization steps in order to continuously tune the architecture and geometry on the basis of the need for handling and observation. The only way to meet these needs and allow the best experimental practice is the use of easy controllable methods of fabrication and low-cost materials. In our research activity, we optimized the realization of microfluidic tools, from the CAD design to the LOC application. In particular, we explored the possibility to obtain monolithic 3D structures with buried running channels, starting from separately microfabricated PMMA layers. In a previous study, we introduced an approach to fabricating interconnected multilevel channels for passive separation. In that work, we discussed the realization of a multilevel/3D device by using layers of LOR and SU-8 photoresists in combination with PDMS. The device, designed for biological applications, consists of two vertically stacked U-shaped channels, which share a central segment obtained by a multistep optical lithography process including a sacrificial layer modified by chemical reactions [40]. Motivated by further improving multilayer stability and fabrication processes and using materials suitable also for industrial exploitation [41], herein we demonstrate and test a novel, reusable 3D microfluidic device, in which the key characteristic is the inclusion of microfabricated PMMA substrates to obtain a structure that can work as an efficient micromixer or a gradient maker tool for LOC integration. The selection of PMMA as a rough material for LOC development derives from its suitability for milling molding and for biological applications, due to its mechanical and chemical characteristics, e.g., good thermal stability, transparency, chemical inertness, and biocompatibility. The multilayer structure has been assembled through a facile and low-cost solvent-assisted method. The obtained device showed increasing complexity and has been demonstrated to work with two different functions: mixing and obtaining of gradients into microchannels, which would take advantage of all the properties that 3D PMMA can offer. The realization of this technology expands on also the possibility of real-time observation, thanks to the complete transparency of the overlapped layers. Further improvements and envisioned applications examples would be cell labeling, a mix of two different kinds of cells in a specific ratio, or even dilution of cells for counting.

Supplementary Materials: The following are available online at https://www.mdpi.com/2072-666X/12/2/104/s1, Video S1: "Particle mixing"; Video S2: "Gradient generator".

Author Contributions: Conceptualization, F.F.; methodology, F.F. and M.S.C.; simulations, S.Z. and F.F.; investigation, S.Z., M.S.C. and F.F.; optimization of experimental conditions, I.T.; data curation, S.Z. and M.S.C.; writing of the original draft preparation, S.Z.; writing of review and editing, M.S.C. and F.F.; supervision, F.F.; project administration, M.S.C. and F.F.; funding acquisition, F.F. and M.S.C. All authors have read and agreed to the published version of the manuscript.

Funding: This research was supported by the following projects: "SMILE (SAW-MIP Integrated device for oraL cancer Early detection) project, part of the ATTRACT programme funded by European

Union's Horizon 2020 Research and Innovation program (grant agreement: 777222)", PRIN 2017 Project—"Prostate cancer: disentangling the relationships with tumor microenvironment to better model and target tumor progression" (grant agreement: Prot. 20174PLLYN), and "TecnoMed Project ("Tecnopolo per la medicina di Precisione—Regione Puglia", grant number: CUP B84I18000540002).

Conflicts of Interest: The authors declare no conflict of interest. The funders had no role in the design of the study; in the collection, analyses, or interpretation of data; in the writing of the manuscript, or in the decision to publish the results.

References

1. Sia, S.K.; Whitesides, G.M. Microfluidic devices fabricated in poly(dimethylsiloxane) for biological studies. *Electrophoresis* **2003**, *24*, 3563–3576. [CrossRef] [PubMed]
2. Wu, H.K.; Huang, B.; Zare, R.N. Construction of microfluidic chips using polydimethylsiloxane for adhesive bonding. *Lab Chip* **2005**, *5*, 1393–1398. [CrossRef] [PubMed]
3. Xu, B.J.; Liu, Z.B.; Lee, Y.K.; Mak, A.; Yang, M. A PDMS microfluidic system with poly(ethylene glycol)/SU-8 based apertures for planar whole cell-patch-clamp recordings. *Sens. Actuator A Phys.* **2011**, *166*, 219–225. [CrossRef]
4. Jia, X.Y.; Che, B.C.; Jing, G.Y.; Zhang, C. Air-Bubble Induced Mixing: A Fluidic Mixer Chip. *Micromachines* **2020**, *11*, 195. [CrossRef] [PubMed]
5. Van Midwoud, P.M.; Janse, A.; Merema, M.T.; Groothuis, G.M.M.; Verpoorte, E. Comparison of Biocompatibility and Adsorption Properties of Different Plastics for Advanced Microfluidic Cell and Tissue Culture Models. *Anal. Chem.* **2012**, *84*, 3938–3944. [CrossRef] [PubMed]
6. Mohammed, M.I.; Haswell, S.; Gibson, I. Lab-on-a-chip or Chip-in-a-lab: Challenges of Commercialization Lost in Translation. *Procedia Technol.* **2015**, *20*, 54–59. [CrossRef]
7. Temiz, Y.; Lovchik, R.D.; Kaigala, G.V.; Delamarche, E. Lab-on-a-chip devices: How to close and plug the lab? *Microelectron. Eng.* **2015**, *132*, 156–175. [CrossRef]
8. Rotem, A.; Abate, A.R.; Utada, A.S.; Van Steijn, V.; Weitz, D.A. Drop formation in non-planar microfluidic devices. *Lab Chip* **2012**, *12*, 4263–4268. [CrossRef]
9. Takao, H.; Miyamura, K.; Ebi, H.; Ashiki, M.; Sawada, K.; Ishida, M. A MEMS microvalve with PDMS diaphragm and two-chamber configuration of thermo-pneumatic actuator for integrated blood test system on silicon. *Sens. Actuator A Phys.* **2005**, *119*, 468–475. [CrossRef]
10. Ghazal, A.; Lafleur, J.P.; Mortensen, K.; Kutter, J.P.; Arleth, L.; Jensen, G.V. Recent advances in X-ray compatible microfluidics for applications in soft materials and life sciences. *Lab Chip* **2016**, *16*, 4263–4295. [CrossRef]
11. Berthier, E.; Young, E.W.K.; Beebe, D. Engineers are from PDMS-land, Biologists are from Polystyrenia. *Lab Chip* **2012**, *12*, 1224–1237. [CrossRef] [PubMed]
12. Huang, B.; Wu, H.K.; Kim, S.; Zare, R.N. Coating of poly(dimethylsiloxane) with n-dodecyl-beta-D-maltoside to minimize nonspecific protein adsorption. *Lab Chip* **2005**, *5*, 1005–1007. [CrossRef] [PubMed]
13. Chumbimuni-Torres, K.Y.; Coronado, R.E.; Mfuh, A.M.; Castro-Guerrero, C.; Silva, M.F.; Negrete, G.R.; Bizios, R.; Garcia, C.D. Adsorption of proteins to thin-films of PDMS and its effect on the adhesion of human endothelial cells. *RSC Adv.* **2011**, *1*, 706–714. [CrossRef] [PubMed]
14. Yu, X.L.; Xiao, J.Z.; Dang, F.Q. Surface Modification of Poly(dimethylsiloxane) Using Ionic Complementary Peptides to Minimize Nonspecific Protein Adsorption. *Langmuir* **2015**, *31*, 5891–5898. [CrossRef] [PubMed]
15. Nguyen, T.; Chidambara, V.A.; Andreasen, S.Z.; Golabi, M.; Huynh, V.N.; Linh, Q.T.; Bang, D.D.; Wolff, A. Point-of-care devices for pathogen detections: The three most important factors to realise towards commercialization. *Trends Anal. Chem.* **2020**, *131*. [CrossRef]
16. Izadi, D.; Nguyen, T.; Lapidus, L.J. Complete Procedure for Fabrication of a Fused Silica Ultrarapid Microfluidic Mixer Used in Biophysical Measurements. *Micromachines* **2017**, *8*, 16. [CrossRef]
17. Volpe, A.; Krishnan, U.; Chiriacò, M.S.; Primiceri, E.; Ancona, A.; Ferrara, F. A smart procedure for the femtosecond laser-based fabrication of a polymeric lab-on-a-chip for capturing tumor cell. *Engineering* **2020**. [CrossRef]
18. Vladisavljević, G.T.; Kobayashi, I.; Nakajima, M. Production of uniform droplets using membrane, microchannel and microfluidic emulsification devices. *Microfluid. Nanofluid.* **2012**, *13*, 151–178. [CrossRef]
19. Shivhare, P.K.; Bhadra, A.; Sajeesh, P.; Prabhakar, A.; Sen, A.K. Hydrodynamic focusing and interdistance control of particle-laden flow for microflow cytometry. *Microfluid. Nanofluid.* **2016**, *20*, 14. [CrossRef]
20. Chicharo, A.; Martins, M.; Barnsley, L.C.; Taouallah, A.; Fernandes, J.; Silva, B.F.B.; Cardoso, S.; Dieguez, L.; Espina, B.; Freitas, P.P. Enhanced magnetic microcytometer with 3D flow focusing for cell enumeration. *Lab Chip* **2018**, *18*, 2593–2603. [CrossRef]
21. Paie, P.; Bragheri, F.; Di Carlo, D.; Osellame, R. Particle focusing by 3D inertial microfluidics. *Microsyst. Nanoeng.* **2017**, *3*. [CrossRef]
22. Mukherjee, P.; Wang, X.; Zhou, J.; Papautsky, I. Single stream inertial focusing in low aspect-ratio triangular microchannels. *Lab Chip* **2019**, *19*, 147–157. [CrossRef] [PubMed]
23. Shallan, A.I.; Smejkal, P.; Corban, M.; Guijt, R.M.; Breadmore, M.C. Cost-Effective Three-Dimensional Printing of Visibly Transparent Microchips within Minutes. *Anal. Chem.* **2014**, *86*, 3124–3130. [CrossRef] [PubMed]

24. Pugliese, M.; Ferrara, F.; Bramanti, A.P.; Gigli, G.; Maiorano, V. In-plane cost-effective magnetically actuated valve for microfluidic applications. *Smart Mater. Struct.* **2017**, *26*. [CrossRef]

25. Lu, Y.; Zhang, M.L.; Zhang, H.X.; Huang, J.Z.; Wang, Z.; Yun, Z.L.; Wang, Y.Y.; Pang, W.; Duan, X.X.; Zhang, H. On-chip acoustic mixer integration of electro-microfluidics towards in-situ and efficient mixing in droplets. *Microfluid. Nanofluid.* **2018**, *22*. [CrossRef]

26. Nasiri, R.; Shamloo, A.; Akbari, J.; Tebon, P.; Dokmeci, M.R.; Ahadian, S. Design and Simulation of an Integrated Centrifugal Microfluidic Device for CTCs Separation and Cell Lysis. *Micromachines* **2020**, *11*, 699. [CrossRef] [PubMed]

27. Rasouli, M.; Mehrizi, A.A.; Goharimanesh, M.; Lashkaripour, A.; Bazaz, S.R. Multi-criteria optimization of curved and baffle-embedded micromixers for bio-applications. *Chem. Eng. Process. Process Intensif.* **2018**, *132*, 175–186. [CrossRef]

28. Baccouche, A.; Okumura, S.; Sieskind, R.; Henry, E.; Aubert-Kato, N.; Bredeche, N.; Bartolo, J.F.; Taly, V.; Rondelez, Y.; Fujii, T.; et al. Massively parallel and multiparameter titration of biochemical assays with droplet microfluidics. *Nat. Protoc.* **2017**, *12*, 1912–1932. [CrossRef]

29. Stroock, A.D.; Dertinger, S.K.; Ajdari, A.; Mezic, I.; Stone, H.A.; Whitesides, G.M. Chaotic mixer for microchannels. *Science* **2002**, *295*, 647–651. [CrossRef]

30. Cai, G.; Xue, L.; Zhang, H.; Lin, J. A Review on Micromixers. *Micromachines* **2017**, *8*, 274. [CrossRef]

31. Piper, M.; Mueller, A.C.; Karam, S.D. The interplay between cancer associated fibroblasts and immune cells in the context of radiation therapy. *Mol. Carcinog.* **2020**, *59*, 754–765. [CrossRef] [PubMed]

32. Wilson, R.E.; O'Connor, R.; Gallops, C.E.; Kwizera, E.A.; Noroozi, B.; Morshed, B.I.; Wang, Y.M.; Huang, X.H. Immunomagnetic Capture and Multiplexed Surface Marker Detection of Circulating Tumor Cells with Magnetic Multicolor Surface-Enhanced Raman Scattering Nanotags. *ACS Appl. Mater. Interfaces* **2020**, *12*, 47220–47232. [CrossRef] [PubMed]

33. Shang, L.R.; Cheng, Y.; Zhao, Y.J. Emerging Droplet Microfluidics. *Chem. Rev.* **2017**, *117*, 7964–8040. [CrossRef]

34. Vergaro, V.; Carlucci, C.; Cascione, M.; Lorusso, C.; Conciauro, F.; Scremin, B.F.; Congedo, P.M.; Cannazza, G.; Citti, C.; Ciccarella, G. Interaction between Human Serum Albumin and Different Anatase TiO2 Nanoparticles: A Nano-bio Interface Study. *Nanomater. Nanotechnol.* **2015**, *5*, 12. [CrossRef]

35. Sweryda-Krawiec, B.; Devaraj, H.; Jacob, G.; Hickman, J.J. A New Interpretation of Serum Albumin Surface Passivation. *Langmuir* **2004**, *20*, 2054–2056. [CrossRef] [PubMed]

36. Bialopiotrowicz, T.; Janczuk, B. Wettability and surface free energy of bovine serum albumin films. *J. Surfactants Deterg.* **2001**, *4*, 287–292. [CrossRef]

37. Cereda, M.; Cocci, A.; Cucchi, D.; Raia, L.; Pirola, D.; Bruno, L.; Ferrari, P.; Pavanati, V.; Calisti, G.; Ferrara, F.; et al. Q3: A Compact Device for Quick, High Precision qPCR. *Sensors* **2018**, *18*, 19. [CrossRef] [PubMed]

38. Chiriaco, M.S.; Bianco, M.; Nigro, A.; Primiceri, E.; Ferrara, F.; Romano, A.; Quattrini, A.; Furlan, R.; Arima, V.; Maruccio, G. Lab-on-Chip for Exosomes and Microvesicles Detection and Characterization. *Sensors* **2018**, *18*, 3175. [CrossRef]

39. Chiriacò, M.S.; Rizzato, S.; Primiceri, E.; Spagnolo, S.; Monteduro, A.G.; Ferrara, F.; Maruccio, G. Optimization of SAW and EIS sensors suitable for environmental particulate monitoring. *Microelectron. Eng.* **2018**. [CrossRef]

40. Chiriaco, M.S.; Bianco, M.; Amato, F.; Primiceri, E.; Ferrara, F.; Arima, V.; Maruccio, G. Fabrication of interconnected multilevel channels in a monolithic SU-8 structure using a LOR sacrificial layer. *Microelectron. Eng.* **2016**, *164*, 30–35. [CrossRef]

41. Parekh, M.; Ali, A.; Ali, Z.; Bateson, S.; Abugchem, F.; Pybus, L.; Lennon, C. Microbioreactor for lower cost and faster optimisation of protein production. *Analyst* **2020**, *145*, 6148–6161. [CrossRef] [PubMed]

micromachines

Article

Cell Sorting Using Electrokinetic Deterministic Lateral Displacement

Bao D. Ho, Jason P. Beech and Jonas O. Tegenfeldt *

Division of Solid State Physics and NanoLund, Physics Department, Lund University, PO Box 118,
22100 Lund, Sweden; bao.hodang@gmail.com (B.D.H.); jason.beech@ftf.lth.se (J.P.B.)
* Correspondence: jonas.tegenfeldt@ftf.lth.se; Tel.: +46-46-222-8063

Abstract: We show that by combining deterministic lateral displacement (DLD) with electrokinetics, it is possible to sort cells based on differences in their membrane and/or internal structures. Using heat to deactivate cells, which change their viability and structure, we then demonstrate sorting of a mixture of viable and non-viable cells for two different cell types. For *Escherichia coli*, the size change due to deactivation is insufficient to allow size-based DLD separation. Our method instead leverages the considerable change in zeta potential to achieve separation at low frequency. Conversely, for *Saccharomyces cerevisiae* (Baker's yeast) the heat treatment does not result in any significant change of zeta potential. Instead, we perform the sorting at higher frequency and utilize what we believe is a change in dielectrophoretic mobility for the separation. We expect our work to form a basis for the development of simple, low-cost, continuous label-free methods that can separate cells and bioparticles based on their intrinsic properties.

Keywords: electrokinetic deterministic lateral displacement; charge-based separation; dielectrophoresis

1. Introduction

While standard cell sorting schemes rely on labelling and molecular recognition events, cells have important properties for which there exist no labels. This has driven the development of microfluidics-based label-free techniques that exploit cells' intrinsic physical properties for fractionation. The targeted properties can be size [1–5], shape [6,7], compressibility [4], dielectric properties [8], or any other physical characteristic, which can be used as a handle to apply a separating force.

Dielectric and electrokinetic properties of the cells can be strongly affected by changes of the structure of the cells [9] (see Figure 1). Dielectrophoresis (DEP) is a well-established technique that can be used to target these types of changes. It is based on the movement of particles along an electric field gradient due to their dielectric properties [10,11]. The sign of the force depends on the difference between the polarizability of the particle and its surrounding medium. The particles may thus experience a force towards higher electric fields (positive DEP) or towards lower electric fields (negative DEP). Adjusting the frequency of the applied electric field, the force can be tuned and made sensitive to the desired types of changes of the particles of interest. One common way to change the structure of cells is heat treatment. It is known to impart structural changes to the membrane and cell wall as well as internal components affecting the viability of yeast [12,13] and bacteria [14,15], with important implications for the food industry. DEP has, therefore, been found useful to characterize heat-treated cells with respect to viability. Pohl and Hawk published a pioneering paper in 1966 showing the ability to separate live yeast cells from stained dead cells using DEP with a simple device consisting of two macroelectrodes inside a PMMA (polymethyl methacrylate) fluid chamber [16]. Markx et al. showed similar results but with more sophisticated devices featuring castellated microelectrodes deposited at the bottom of a microfluidic chamber [17,18]. By adjusting the frequency of the AC voltage applied between adjacent electrodes and switching the flow and voltages, the authors

Citation: Ho, B.D.; Beech, J.P.;
Tegenfeldt, J.O. Cell Sorting Using
Electrokinetic Deterministic Lateral
Displacement. *Micromachines* **2021**, 12,
30. https://doi.org/ 10.3390/
mi12010030

Received: 10 December 2020
Accepted: 24 December 2020
Published: 30 December 2020

Publisher's Note: MDPI stays neutral with regard to jurisdictional claims in published maps and institutional affiliations.

implemented a trap and release separation scheme. Trapping of viable and non-viable *Escherichia coli* bacteria using DEP in insulating array structures has also been reported [19]. Here, electroosmosis was used for flow, and live/dead bacteria were trapped at different positions in the insulating array using DC-DEP and a difference in their DEP mobility. Not only DEP but also other types of electrokinetic forces can be leveraged for particle sorting. We showed recently, how zeta potential can be used as a basis for sorting of particles of different surface charge [20].

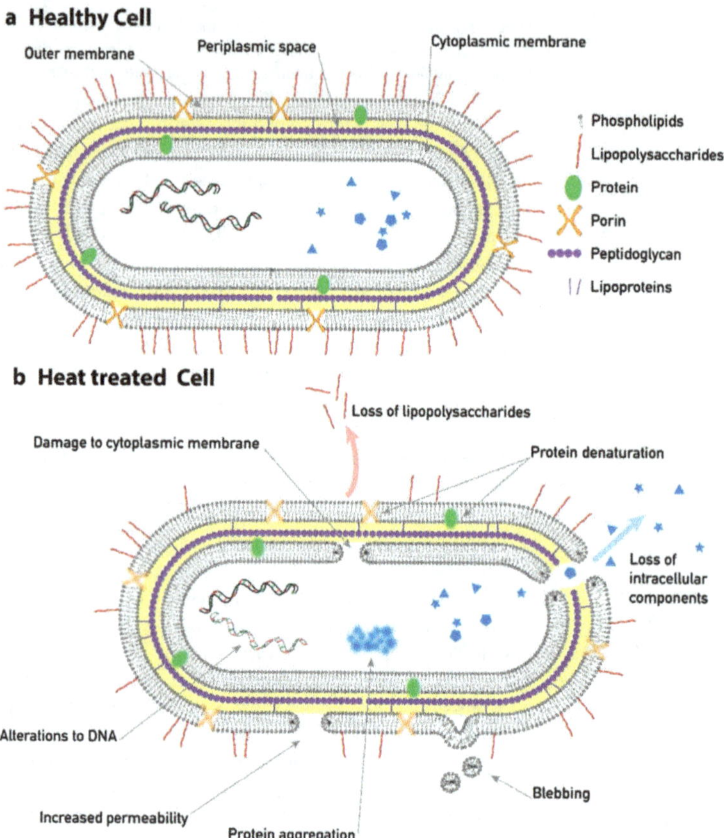

Figure 1. An overview of typical structural changes that occur as a consequence of heat treatment. Here, the cell is a gram-negative bacteria, but similar mechanisms occur in other cell types. Structural changes of a cell are reflected in its dielectric and electrokinetic properties due to, e.g., changes in the conductivity of the membrane, the charge of membrane, and aggregation of protein. Schematics are based on descriptions of heat-induced changes in [9,15].

In this work, we will use untreated and heat-treated *Escherichia coli* and *Saccharomyces cerevisiae* as model systems for cells with different internal structure to demonstrate the capabilities of a combination of electrokinetics and deterministic lateral displacement (DLD) to sort cells.

DLD is a powerful mechanism for sorting particles based on size [2]. Among the strengths of DLD are that it is continuous, making it suitable for integration with other methods (see the CTC-iChip [21]), it has high size resolution [2], and has been demonstrated to work for a wide range of particle sizes, from millimeters [22] to nanometers [23]. As a result, DLD has been employed in various microfluidic devices to sort a wide range of cells and bioparticles, i.e., white blood cells from red blood cells and plasma [24–26], circulating tumor cells from blood [27,28], and trypanosomes from blood [29,30].

The basic mechanism of DLD is described in Figure 2. Particles move through the array in one out of two basic modes governed by steric interactions with the posts in the array, small particles following the flow in the zigzag mode and large particles deviated laterally according to the array geometry in the displacement mode. The threshold diameter between the two modes is called the critical diameter, which can be estimated by an empirical formula presented by Davis [31]:

$$D_C = 1.4GN^{-0.48}. \tag{1}$$

Here, G is the gap between adjacent pillars and N is the period of the pillar array, the number of rows along the flow direction after which the array repeats itself.

Through additional force interactions between the posts and the particles, the device can be made more versatile. Electrostatic interactions between particles and posts have been exploited as a means to bias the DLD and make it sensitive to the charge of particles [32]. Another approach is to add electric fields to the device so that electrokinetic forces affect the sorting, opening up for tunability and more specific sorting [33]. The approach is known as electrokinetic DLD (eDLD), and we will apply it here as described recently [20].

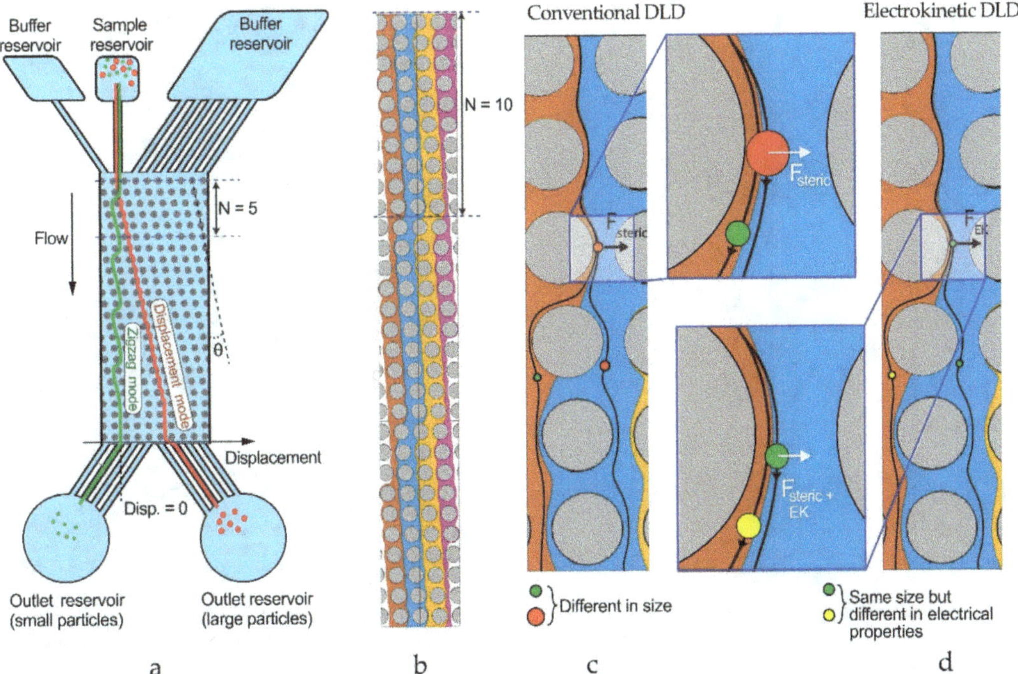

Figure 2. Working principle of deterministic lateral displacement (DLD). (**a**) An array of pillars tilted with an angle splits a mixture of particles into different trajectories based on size. At the end of the array, the different-sized particles (green and red) exit at different lateral positions and are collected into different outlet reservoirs. (**b**) Simulated flow streams within a DLD array, showing zigzagging patterns. (**c**) Mechanism of sorting: steric interactions between particles and posts cause particles to change flow streams in a size-dependent manner that leads to separation. (**d**) In an electrokinetic DLD device, electrokinetic forces act on the particles in addition to the steric forces, modifying separations.

2. Materials and Methods

2.1. Devices and Experimental Setup

The DLD devices were cast in polydimethylsiloxane (PDMS) using replica molding [34]. Details can be found in Table S1, Supplementary Information. We have used two types of designs: Analytical devices (Figure 3a and Table 1) and Sorting devices (Figure 3b and Table 2). In an Analytical device, only a narrow, single stream of sample is allowed to enter the DLD array and is buffered on two sides. The particles in the sample are separated as they travel along the array before entering a single outlet reservoir. The lateral positions of the particles are analyzed at the observation windows at the beginning and at the end of the array. In the second type of device, the Sorting device, the input sample stream is much wider to increase throughput. At the outlet side, there are several output reservoirs collecting particles that are displaced by different amounts and that can be counted externally.

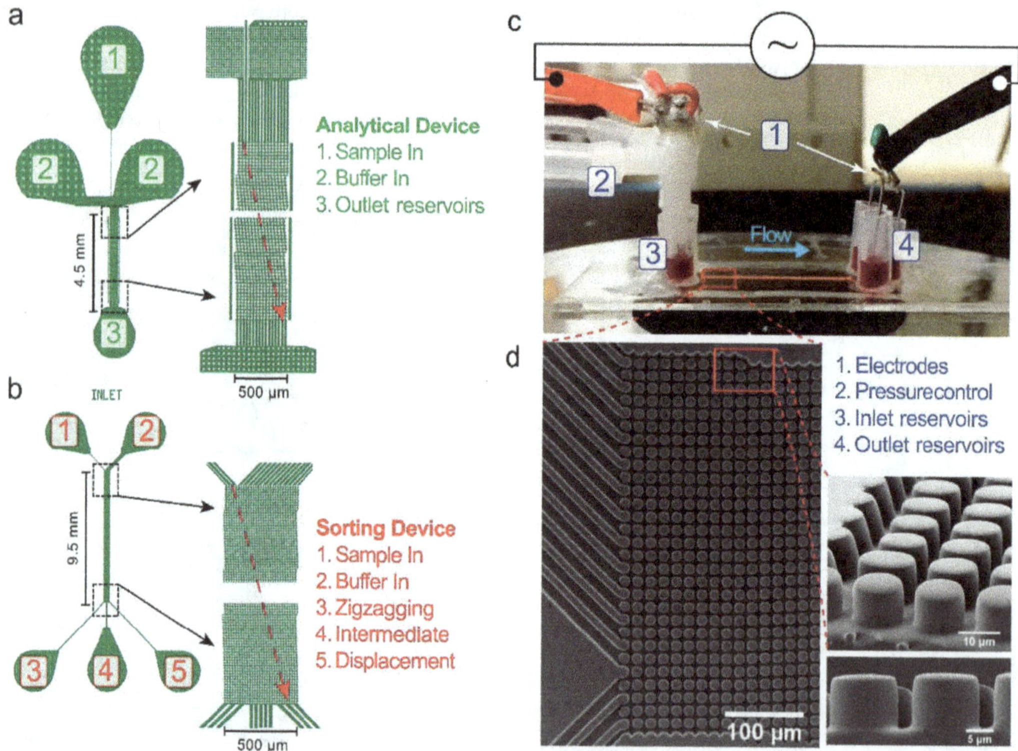

Figure 3. Details of the devices and the experimental setup. (**a**,**b**) Two types of designs used. (**c**) Side view of a device and the experimental setup, with pressure and electrical connections. (**d**) SEM images of the DLD array of Sorting device #1, which is representative for all devices used in this work.

Table 1. Analytical deterministic lateral displacement (DLD) devices. Critical diameters, D_C, are nominal values given based on the geometry of the devices and Equation (1). Channel lengths are 4.25 ± 0.25 mm. The deflection width is approximately 425 µm corresponding to 85% of the device width.

Device Name	Gap (µm)	N	D_C (µm)	Deflection, θ	Cell Type
Analytical device #1	4	10	1.9	5.71°	*E. coli*
Analytical device #2	10	10	4.6	5.71°	Yeast
Analytical device #3	11	10	5.1	5.71°	Yeast
Analytical device #4	12	10	5.6	5.71°	Yeast

Table 2. Sorting devices used for *E. coli*. The trajectories for the large particles (displacement mode) are illustrated through four different numbers: critical diameter, deflection angle, absolute deflection, and relative deflection. Critical diameters, D_C, are nominal values given based on the geometry of the devices and Equation (1).

Device Name	Gap (µm)	N	D_C (µm)	Channel Length (mm)	Deflection θ	Deflection (µm)	Deflection/Channel Width
Sorting device #1	4	23	1.24	9.5	2.49°	400	86%
Sorting device #2	3	50	0.64	22.9	1.15°	450	86%

The setup of an experiment is shown in Figure 3c. The device was placed on the stage of an inverted microscope (Nikon Eclipse TE2000-U, Nikon Corporation, Tokyo, Japan). At inlet reservoirs, an overpressure (1–100 mBar) was applied via a pressure controller (MFCS-4C, Fluigent, Paris, France). Together with the pressure, a voltage was applied between inlet and outlet reservoirs via platinum electrodes in the reservoirs. A function generator (33120A, Hewlett Packard, Palo Alto, CA, USA), in combination with a high-voltage amplifier (Bipolar Operational Power Supply/Amplifier BOP 1000M, Kepco, Flushing, NY, USA), or a high-frequency amplifier (WMA-300, Falco Systems, Amsterdam, The Netherlands) were used to provide the required signals. The voltage was confirmed with an oscilloscope (54603B 60 MHz, Hewlett Packard, Palo Alto, CA, USA) via a 1x/10x probe (Kenwood PC-54, 600 Vpp, Havant, UK). The experimental image stacks were captured with a monochrome Andor Neo sCMOS camera (Andor Technology, Belfast, Northern Ireland) or a color camera (Exmor USB 3.0, USB29 UXG M, Sony, Tokyo, Japan). The stacks were then analyzed using FIJI (ImageJ 1.52f, National Institutes of Health, Bethesda, MD, USA) and MATLAB (MathWorks, Natick, MA, USA). To measure the conductivity of the media, we used a conductivity meter (B-771 LAQUAtwin, Horiba Instruments, Kyoto, Japan). The zeta potentials of cells were measured with a Zetasizer NanoZS instrument (Malvern Instruments, Ltd., Worcestershire, UK).

2.2. Data Analysis

For both *S. cerevisiae* and *E. coli*, sorting performance during an experiment is assessed by comparison of lateral positions of cell populations at the beginning and at the end of the DLD array. The degree of displacement is given as the number of gaps from one side to the other at the end of the device. Particles that follow the flow exit at low gap numbers, and particles that are displaced exit at high gap numbers. However, when plotting the results, we convert the gap numbers to percentages (0%: no displacement, 100%: maximal displacement) as this makes it easier to compare results from devices with different numbers of gaps. The particles are counted in two ways. Manual counting, cell by cell, is accurate but can be labor intensive. This method was performed for the yeast cells, which are nonfluorescent. For the cells that are fluorescent (viable/non-viable *E. coli*), the fluorescence intensity is used as well to deduce the numbers of particles, which we will refer to in the plots as inferred counts. More details of the image processing can be found in Figure S1 of the Supplementary Information.

In the experiments with *E. coli* where the bacteria were sorted into different outlet reservoirs, the numbers of viable and non-viable cells recovered from the sample and the outlet reservoirs were evaluated. First, the recovered suspension from each outlet reservoir was pipetted into a centrifuge tube and concentrated by centrifugation. After that, the concentrated suspension was pipetted on a microscope slide, covered, and sealed with a cover slip. The cells were then imaged and counted to give the ratio of viable to non-viable cells.

2.3. Sample Preparation

We use bacteria and yeast cells as model systems to demonstrate proof of principle of our devices. Key information about the cells can be found in Section 3 of the Supplementary Information.

Green fluorescent *Escherichia coli* (2566/pGFP) (approximately 1.5 μm × 3 μm) were cultured and stored at −80 °C in culture medium with 20% w/v glycerol. Prior to experiments, the bacteria were allowed to thaw at room temperature. The concentration of the bacteria was measured at 8.2×10^8/mL, using a DMS cell density meter with 600 nm light (DMS-cuvette, LAXCO Inc., Bothell, WA, USA). The bacteria were then spun down and suspended in running medium (KCl + 0.1% w/v Pluronic® F127, σ = 20, 100, or 500 mS/m) at the same concentration. Half of the sample was kept at 70 °C for 20 min to kill the cells and then both halves were stained with propidium iodide (PI) (Sigma-Aldrich Sweden AB, Stockholm, Sweden) at a concentration of 20 μg/mL for 5 min. Propidium iodide is a dead-cell stain, which can penetrate compromised cell membranes. Viable *E. coli* bacteria will appear green due to the GFP, prior to and after PI staining. The heat-treated *E. coli* bacteria will appear dark prior to PI staining and orange after PI staining (using our microscope setup), enabling viable and non-viable heat-treated cells to be distinguished. The staining revealed that, prior to running experiments, in the "viable" population, around 80–90% of the cells were actually viable and in the heat-treated "non-viable" population, around 90–95% were actually non-viable.

Baker's yeast cells (*Saccharomyces cerevisiae*) ($D \sim 4.5$ μm) in dry form (Jästbolaget AB, Sollentuna, Sweden) were suspended in glucose 5% w/v and heated up to 32 °C in a well-ventilated tube for 30 min, for activation. Half of the sample was then heat treated at 62 °C for 15 min to kill the cells. The viable and heat-treated samples were mixed at a ratio of 1:1 and stained with Trypan Blue at a concentration of 0.2% w/v, for 5 min. Trypan Blue is a common dead-cell stain, which permeates compromised cell membranes, leaving the non-viable cells dark blue and enabling viable and non-viable cells to be distinguished from one another. The mixed sample was then washed several times with the running media, consisting of KCl at different conductivities and 0.1% w/v Pluronic® F127. The staining revealed that, prior to running experiments, in the "viable" population, around 70% of the cells were actually viable and in the heat-treated "non-viable" population, more than 90% were actually non-viable

3. Results and Discussion

We first characterize the different cell types with respect to size and zeta potential. See Table S2 in the Supplementary Information. Different experimental parameters are then explored by testing devices with different critical sizes, and different combinations of applied pressure, applied voltage, frequency of the applied voltage and conductivity of the buffer. Finally, we show that we can sort the *E. coli* with respect to zeta potential and, by selecting slightly different experimental conditions, the yeast based on what we believe is DEP. Estimated throughputs and Péclet numbers are given in Table S3 of the Supplementary Information.

3.1. Sorting of Viable/Non-Viable E. coli

We measured the size of rod-shaped *E. coli* (~2.5 μm × 1.5 μm, Figure S2, Supplementary Information) and found that the size differences between viable and non-viable bacteria are negligible. It is, therefore, not possible to separate them by size. We could, however, measure a significant change in the zeta potential (approximately −42 and −34 mV for viable and non-viable bacteria, respectively) using the Zetasizer™.

To find optimal experimental conditions for sorting the two different cell states, we started by selecting a suitable device. Sorting device #2 (D_C = 0.64 μm) was found to have a too small critical diameter to function (Figure S3, Supplementary Information). Viable and non-viable cells had overlapping distributions at the end of the device and both were in displacement mode even when no voltage was applied. With an applied voltage at 1 to 10 Hz, displacement increased and overlap remained.

We, therefore, shifted to Sorting device #1 (D_C = 1.24 μm), where both viable and non-viable cells travelled in zigzag mode when no voltage was applied. Supplementary Information Video S1 and Video S2, show bacteria distributions at the inlet and at the outlet, respectively. As the applied AC voltage was increased from zero, the cells transit from zigzag to displacement mode, and the viable cells tended to displace more easily than the non-viable cells, leading to separation (Supplementary Information Video S3).

The first test now is to scan the frequency from single Hz to hundreds of kHz. Figure 4 demonstrates clearly that viable *E. coli* are not displaced for frequencies above 1 kHz indicating that their zeta potential will determine their trajectories through the device rather than their dielectrophoretic mobility [20,35].

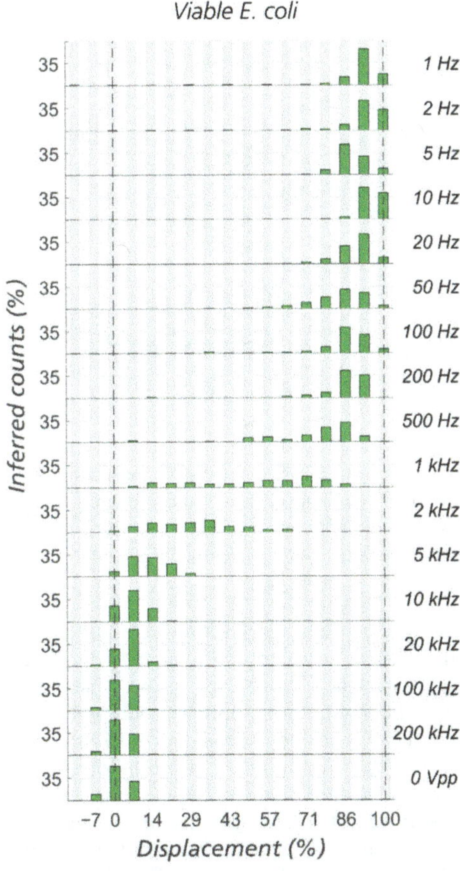

Figure 4. Effect of frequency on displacement of viable *E. coli* in eDLD. Analytical device #1 with gap 4 μm, N = 10, D_C = 1.85 μm, medium conductivity = 25 mS/m, V = 300 V$_{PP}$, and ΔP = 9.5 mBar.

Since the zeta potential is an important sorting parameter, we expect that it is necessary to carefully select buffer conductivity for optimal sorting. Therefore, for three different conductivities, we varied the frequency of the applied electric field in order to find an optimal combination of these two parameters. See Figure 5 for a summary of the data. The given voltages represent the lowest values for which a clear separation could be observed.

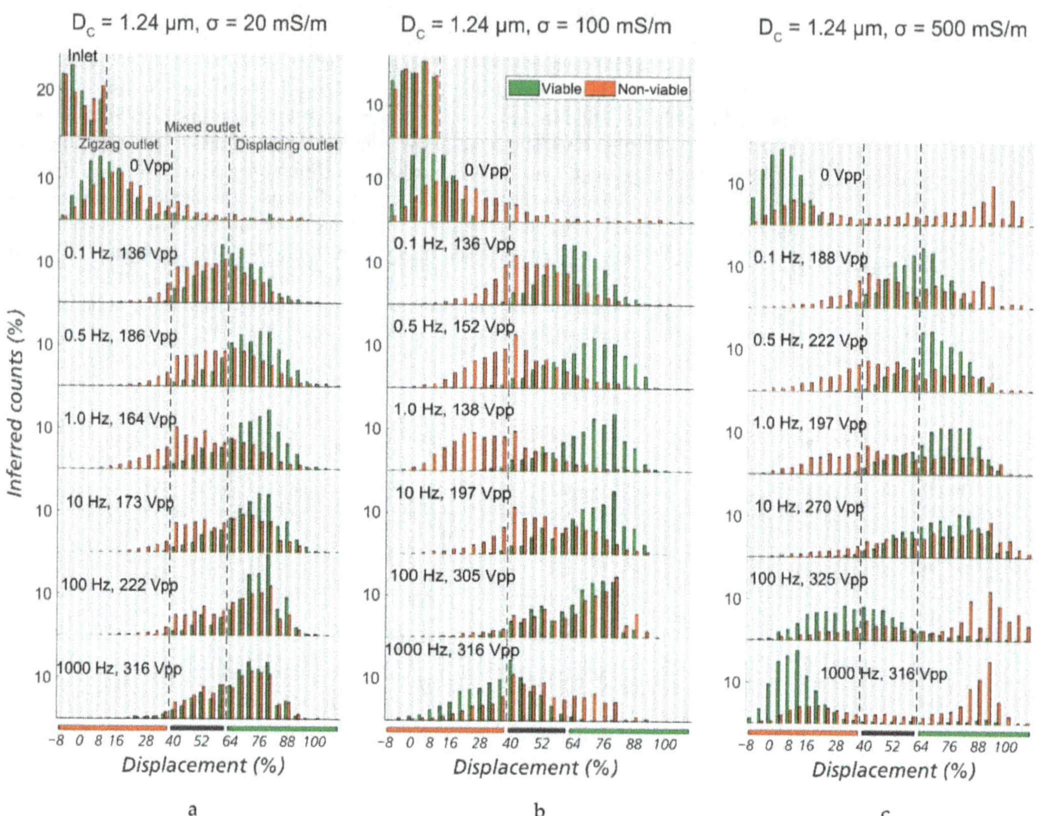

Figure 5. Optimization of sorting conditions. Lateral displacement of viable and non-viable *E. coli* at the outlet of the device (Sorting Device #1, D_C = 1.24 µm), at different frequencies, voltages, and conductivities of the media: (**a**) Medium conductivity of 20 mS/m, (**b**) medium conductivity of 100 mS/m, and (**c**) medium conductivity of 500 mS/m. It can be seen that in general, σ = 100 mS/m gives the best separation among the three. Specifically, the optimized conditions for sorting are σ = 100 mS/m, f = 1.0 Hz, and V = 138 V_{PP}.

As the frequency is increased from 1 Hz, a higher voltage is needed to steer viable cells into the "displacing reservoir" (denoted by the green bar under the plots) and the "intermediate reservoir" (denoted by the grey bar). At 1 kHz, even a voltage of more than 300 V_{PP} (the limit of our amplifier) could not fully displace viable bacteria. It can be seen that 1 Hz was the best frequency for sorting, at a conductivity of 100 mS/m. At conductivities of 20 or 500 mS/m, similar trends were observed but the sorting capability was less efficient than at 100 mS/m. This is consistent with what we see for polystyrene microspheres as well as lipid vesicles in [20]. The increase in zeta potential of the particles due to the lower ionic strength contributes to the improved sorting.

Note the appearance of large aggregates at high conductivities (Figure 5c). We believe that the screening of the particle charge at the high conductivity of 500 mS/m gives rise to the aggregation. Supplementary Information Video S4 and Video S5, show viable and non-viable *E. coli* suspended in KCl at 0.5 mS/m and 500 mS/m, respectively, and support this view. While almost all non-viable *E. coli* (orange) in 0.5 mS/m medium are singles, many of the non-viable bacteria in 500 mS/m medium form large aggregates.

In the end, the optimized combination of conditions for sorting viable and non-viable *E. coli* was found to be a medium conductivity of 100 mS/m, a sinusoidal voltage of 1 Hz/138 V_{PP} in Sorting device #1 (D_C = 1.24 μm) at an applied pressure of 20 mBar.

Using the optimal conditions, a mixture of viable/non-viable *E. coli* in equal proportions was run through the eDLD device (Sorting device #1, D_C = 1.24 μm) for 1 h and 30 min. After sorting, the populations of the bacteria in the three outlet reservoirs were recovered and counted externally. The outlet reservoirs are named "zigzag, intermediate, and displacement," corresponding to the trajectories of the particles they collect. The results for one such experiment are shown in Figure 6, demonstrating high purity of non-viable cells in the "zigzag reservoir" and high purity of viable cells in the "displacing reservoir." To illustrate the reproducibility of results, three additional repetitions were carried out (Figure S4 of the Supplementary Information). We estimated, based on fluorescent intensity (Figure 6a), that 72% of the viable bacteria were recovered in the displacing reservoir, with a purity, based on manual count (Figure 6b), of more than 90%. Likewise, 63% of the non-viable bacteria were recovered in the zigzag reservoir, with a purity of more than 90%. The remaining cells exit at the intermediate reservoir. Note that depending on the application, purity could be increased at the expense of decreased recovery rate or vice versa by varying the design of the device, more specifically the placement of the exit channels and collection reservoirs. In Section 4 of the Supplementary Information, we show a full range of possible purity and recovery rates when varying placements of the zigzag and displacing reservoirs, see Figure S6.

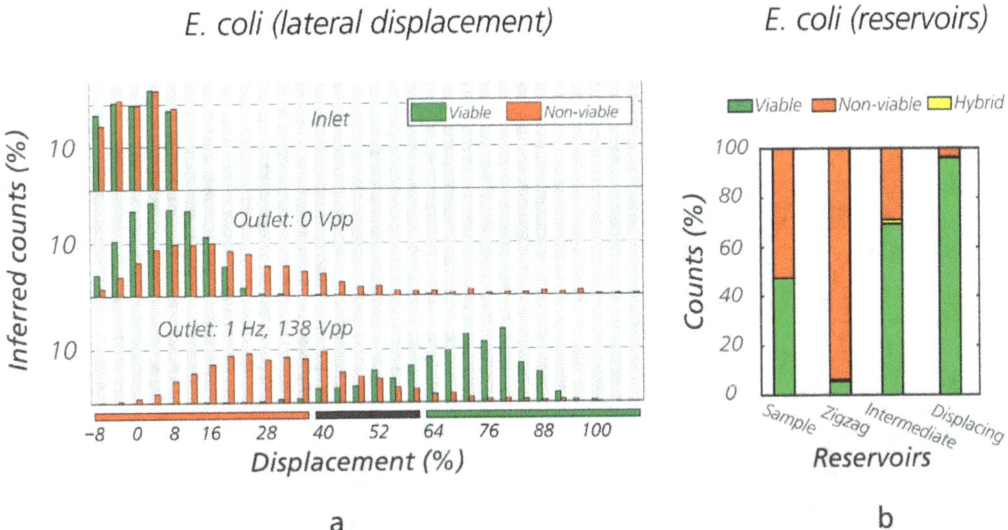

Figure 6. Sorting of viable/non-viable *E. coli* using electrokinetic DLD (eDLD). (**a**) Lateral displacement of the viable and non-viable *E. coli* based on fluorescence intensity at the beginning and at end of a DLD array (Sorting device #1, D_C = 1.24 μm), with and without an applied field. The displacement corresponding to the zigzag/intermediate/displacing reservoirs have been marked with the orange/gray/green horizontal bars, respectively. Note that there seems to be some separation of non-viable cells from viable cells even when the electric field is switched off, probably due to the higher probability of non-viable cells to form aggregates. While roughly half of the non-viable cells can be collected, the remainder is still mixed with viable cells. For the general case, this level of sorting is of limited value. (**b**) External, manual counts of the ratio between viable and non-viable *E. coli* recovered from different outlet reservoirs. "Hybrid" refers to the few cells emitting both green and orange color, which represents a minute subset of the counted cells.

3.2. Sorting of Viable/Non-Viable Yeast Cells

We measured the size of yeast cells (~2.5 μm × 1.5 μm), Figure S5, and found that the size differences between viable and non-viable bacteria is greater than for *E. coli*, but still small relative to the width of the size distributions of both types (non-viable: 3.80 ± 0.44 μm and viable: 4.70 ± 0.63 μm). In contrast to *E. coli*, the measured zeta potential of yeast cells was observed to change very little due to heat inactivation and was not expected to be useful for separation of the two subpopulations (~−19.5 mV, Table S2, Supplementary Information).

We first tested the effects of the frequency and the applied voltage. When changing frequency (Figure 7a), we observed that at low frequency (~100 Hz), yeast cells experience stronger displacement than at higher frequency (1 and 20 kHz), just like polystyrene beads [20] and *E. coli*. When the changing parameter was voltage while the frequency was kept at 100 Hz (Figure 7b), both the viable and the non-viable yeast cells were displaced quite strongly and the displacement monotonically increases with the applied voltage. Separation was, however, not observed at any voltage due to the lack of contrast in electrical properties (zeta potentials) between the viable and the non-viable yeast cells.

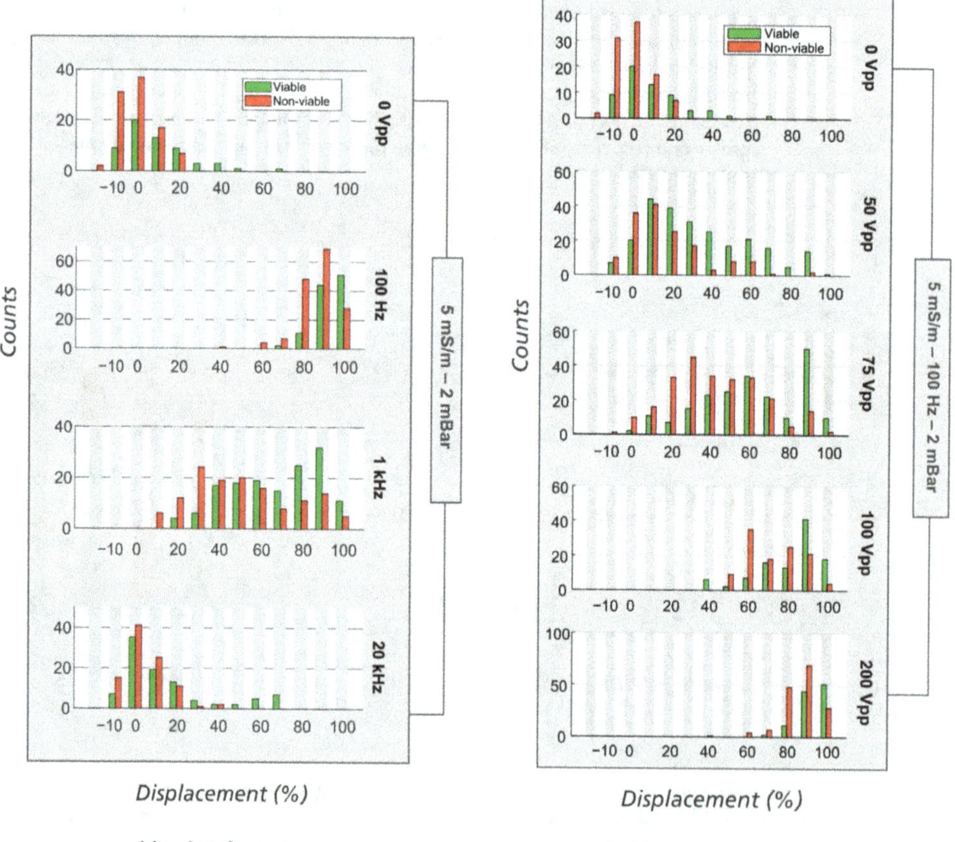

Figure 7. Effect of frequency and voltage on displacement of yeast cells at the outlet of the device. Analytical device #4 (D_C = 5.6 μm) was used with σ = 5 mS/m and ΔP = 2 mBar. (**a**) Varying frequency. Note the trend that the displacement increases equally for both cell types as the frequency is decreased. The applied voltage was 300 V_{PP}, except at 100 Hz, which is 200 V_{PP}. (**b**) Varying voltage. The frequency was kept at 100 Hz.

As in the case of *E. coli*, we then performed experiments at different medium conductivities and in different devices to find the best condition for sorting viable/non-viable yeast. It can be seen in Figure 8a that medium conductivity affects displacement of viable and non-viable yeast cells in eDLD. At the same applied voltage and frequency, some conductivity values give better separation between the viable and the non-viable cells than the others. We found the combination of σ = 50 mS/m, V = 300 V$_{PP}$, f = 20 kHz, and ΔP = 1.5 mBar for Analytical device #2, D_C = 4.60 µm, or ΔP = 1.2 mBar for Analytical device #3, D_C = 5.10 µm, to be optimal for separation of viable/non-viable yeast cells. Such optimal separation is illustrated in Figure 8b. Supporting videos can be found in the Supplementary Information (Video S6: without the electric field and Video S7: with the electric field).

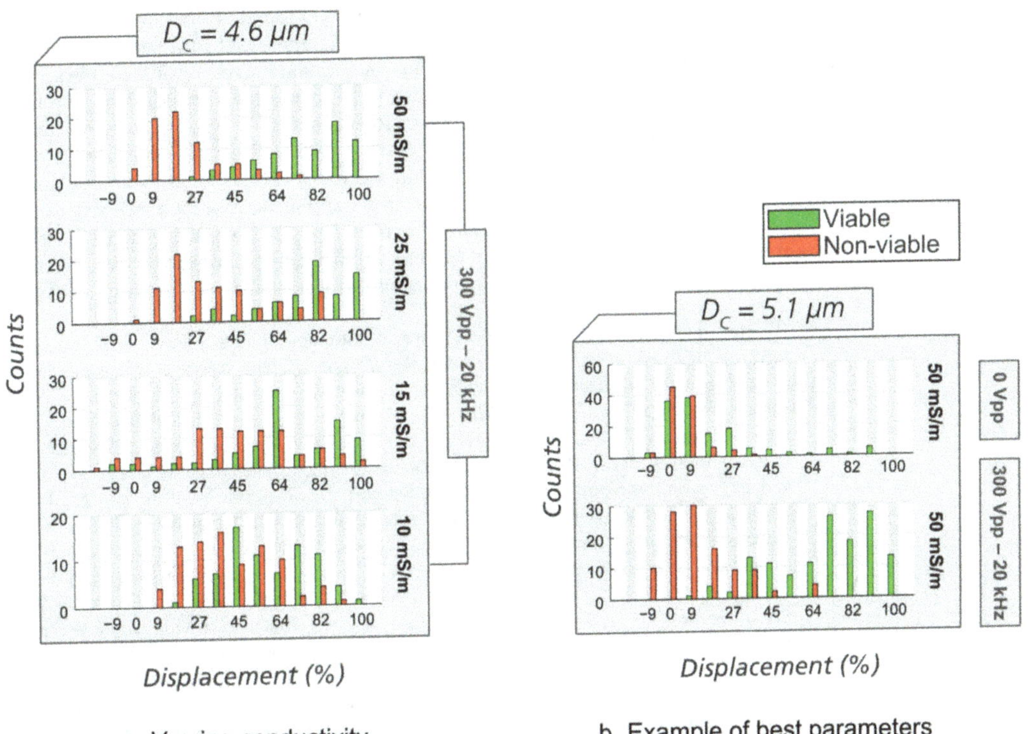

Figure 8. Displacement of viable/non-viable yeast cells at varying medium conductivities and in devices with different D_C. (**a**) Varying conductivity. We use Analytical device #2, D_C = 4.6 µm. In each experiment, the pressures were adjusted to maximize separation (from top to bottom, 1.5, 1.2, 0.7, and 0.6 mBar). (**b**) An example of a set of parameters that gives clear separation (Analytical device #3, D_C = 5.10 µm, σ = 50 mS/m, f = 20 kHz, V = 300 V$_{PP}$, ΔP = 1.2 mBar).

While the low-frequency approach was unsuccessful for yeast, by separating the yeast cells at higher frequency, we demonstrate one of the strengths of our method, namely, frequency can be used to target different types of morphological or structural changes in cells. Since electroosmosis and electrophoresis can be assumed to be negligible at 20 kHz, it is reasonable to assume that DEP makes an important contribution to the enhanced displacement and sorting in this case.

4. Conclusions and Future Work

We have presented and characterized an integrated device that combines DLD and electrokinetics to sort particles based not only on size but also on electric and dielectric properties. Since changes in viability are linked to changes in properties such as surface charge (zeta potential), membrane integrity, membrane conductivity, and polarizability rather than size or shape, we were able to show proof of principle of leveraging these parameters to perform viability-based separations.

We note that the *E. coli* and yeast separations require different ranges of frequency to function. This implies the existence of at least two different mechanisms that make eDLD work: one at low frequency, related to electroosmotic flow (EOF) and electrophoresis (EP) (as we discuss in [20]), and the other at higher frequency, related to DEP. At this point we do not know the exact mechanisms. We can rule out DEP as the main mechanism for the low-frequency cases as well as most probably linear EOF and EP. Instead, we posit that non-linear EOF and EP may play an important role [20]. However, further studies are necessary to fully understand how these mechanisms interact and are responsible for the separation. Nevertheless, we have demonstrated that we can select different underlying mechanisms and optimize running conditions such as ionic strength, applied voltage and pressure, and the geometry of the device, to adapt the approach to two different cells with different physical characteristics that in turn are coupled to relevant biological subpopulations.

Potential applications that we envision are applications in food industry to monitor the effect of the viability of any microorganisms used in the processing of the food and those microorganisms that are not desired. From a fabrication perspective, the design is simple and can be realized in cheap materials using standard mass production techniques.

Supplementary Materials: The following are available online at https://www.mdpi.com/2072-666X/12/1/30/s1. Figure S1: Image processing steps, Figure S2: Dimensions of viable and non-viable *E. coli*, Figure S3: Displacement of viable and non-viable *E. coli* in D_C = 0.64 µm device, Figure S4: Sorting of viable and non-viable *E. coli* in D_C = 1.24 µm device, Figure S5: Dimensions of viable and non-viable yeast cells, Figure S6: Purity and recovery rates of viable and non-viable *E. coli*, Table S1: Device fabrication steps, Table S2: Properties of the cells, Table S3: Throughput and Péclet numbers of various experiments reported in the main text, Video S1: *E. Coli*—Sorting—100 mS_m—20 mBar—Inlet. Viable (green) and non-viable (orange) *E. coli* bacteria entering a DLD array having D_C = 1.24 µm. The video was recorded in real time using the color camera. In the same experiment, the monochrome camera was used to obtain the videos from which the results were processed and plotted using the method described in Figure S1. The results are shown in the top plot of Figure 6a. Video S2: *E. coli*—Sorting—100 mS_m—20 mBar—Outlet. Same experiment as in Video S1, now captured at the outlet. No voltage was applied. The results are shown in the middle plot of Figure 6a. Video S3: *E. coli*—Sorting—100 mS_m—20 mBar—1 Hz—138 Vpp—Outlet. Continuing from Video 2, now with the AC voltage on. Separation of viable and non-viable cells is demonstrated. The results are shown in the bottom plot of Figure 6a. Video S4: *E. coli*—0d5 mS_m. At low medium conductivity, most of the non-viable *E. coli* (orange) stay single. Video S5: *E. coli*—500 mS_m. At high medium conductivity, most of the non-viable *E. coli* (orange) form clusters. Video S6: Yeast—Sorting—50 mS_m—0 Vpp—1d2 mBar. Viable (transparent) and non-viable (dark) yeast cells exiting at the end of a DLD array having D_C = 5.1 µm. Both viable and non-viable cells were in zigzag mode and stay closer to the left wall. The video was cut out from the original video used to produce Figure 8b (0 V_{PP}) and sped up two times. The original video is 10 times longer. Video S7: Yeast—Sorting—50 mS_m—20 kHz—300 Vpp—1d2 mBar. Similar to Video S6, but an AC voltage of 20 kHz, 300 V_{PP} was applied. While the non-viable cells were still in zigzag mode, the viable ones were displaced and appear closer to the right wall. The video was cut out from the original video used to produce the graph of Figure 8b (300 V_{PP}) and sped up two times. The original video is 7.5 times longer.

Author Contributions: Conceptualization, all authors; methodology, all authors; software, B.D.H.; validation, all authors; formal analysis, B.D.H.; investigation, B.D.H.; resources, J.O.T.; data curation, B.D.H.; writing—original draft preparation, B.D.H.; writing—review and editing, all authors; visualization, B.D.H. and J.P.B.; supervision, J.O.T. and J.P.B.; project administration, J.O.T.; funding acquisition, J.O.T. All authors have read and agreed to the published version of the manuscript.

Funding: This research was funded by the European Union, grant number 607350 (project LAPASO), grant number 801367 (project evFOUNDRY), and grant number 634890 (project BeyondSeq); by the Swedish Research council, grant number 2016-05739; and NanoLund. All device processing was conducted within Lund Nano Lab.

Data Availability Statement: Essential data are contained within the article and the supplementary material. The raw data are available on request from the corresponding author.

Acknowledgments: We are thankful to Hywel Morgan, Carlos Honrado, Daniel Spencer, and Nicolas Green (University of Southampton) for their generous sharing of expertise in electrokinetics. We are grateful to Karin Schillén for her help with zeta potential measurements and Lars Hederstedt (Lund University) for his help with the culturing of *E. coli*.

Conflicts of Interest: The authors declare no conflict of interest. The funders had no role in the design of the study, in the collection, analyses, or interpretation of data; in the writing of the manuscript; or in the decision to publish the results.

References

1. Yamada, M.; Nakashima, M.; Seki, M. Pinched Flow Fractionation: Continuous Size Separation of Particles Utilizing a Laminar Flow Profile in a Pinched Microchannel. *Anal. Chem.* **2004**, *76*, 5465–5471. [CrossRef] [PubMed]
2. Huang, L.R.; Cox, E.C.; Austin, R.H.; Sturm, J.C. Continuous Particle Separation Through Deterministic Lateral Displacement. *Science* **2004**, *304*, 987–990. [CrossRef] [PubMed]
3. Di Carlo, D.; Irimia, D.; Tompkins, R.G.; Toner, M. Continuous inertial focusing, ordering, and separation of particles in microchannels. *Proc. Natl. Acad. Sci. USA* **2007**, *104*, 18892–18897. [CrossRef] [PubMed]
4. Petersson, F.; Åberg, L.; Swärd-Nilsson, A.A.-M.; Laurell, T. Free Flow Acoustophoresis: Microfluidic-Based Mode of Particle and Cell Separation. *Anal. Chem.* **2007**, *79*, 5117–5123. [CrossRef]
5. Southern, E.M. Detection of specific sequences among DNA fragments separated by gel electrophoresis. *J. Mol. Biol.* **1975**, *98*, 503–517. [CrossRef]
6. Beech, J.P.; Holm, S.H.; Adolfsson, K.; Tegenfeldt, J.O. Sorting cells by size, shape and deformability. *Lab Chip* **2012**, *12*, 1048–1051. [CrossRef]
7. Beech, J.P.; Ho, B.D.; Garriss, G.; Oliveira, V.; Henriques-Normark, B.; Tegenfeldt, J.O. Separation of pathogenic bacteria by chain length. *Anal. Chim. Acta* **2018**, *1000*, 223–231. [CrossRef]
8. Pethig, R. Review Article—Dielectrophoresis: Status of the theory, technology, and applications. *Biomicrofluidics* **2010**, *4*, 022811. [CrossRef]
9. Flores-Cosío, G.; Herrera-López, E.J.; Arellano-Plaza, M.; Mathis, A.G.; Kirchmayr, M.; Amaya-Delgado, L. Application of dielectric spectroscopy to unravel the physiological state of microorganisms: Current state, prospects and limits. *Appl. Microbiol. Biotechnol.* **2020**, *104*, 6101–6113. [CrossRef]
10. Pohl, H.A. *Dielectrophoresis: The Behavior of Neutral Matter in Nonuniform Electric Fields (Cambridge Monographs on Physics)*; Cambridge University Press: Cambridge, UK, 1978.
11. Morgan, H.; Green, N.G. *AC Electrokinetics: Colloids and Nanoparticles*; Research Studies Press: Hertfordshire, UK, 2003.
12. Guyot, S.; Gervais, P.; Young, M.; Winckler, P.; Dumont, J.; Davey, H.M. Surviving the heat: Heterogeneity of response inSaccharomyces cerevisiaeprovides insight into thermal damage to the membrane. *Environ. Microbiol.* **2015**, *17*, 2982–2992. [CrossRef]
13. Pillet, F.; Lemonier, S.; Schiavone, M.; Formosa, C.; Martin-Yken, H.; François, J.-M.; Dague, E. Uncovering by Atomic Force Microscopy of an original circular structure at the yeast cell surface in response to heat shock. *BMC Biol.* **2014**, *12*, 6. [CrossRef] [PubMed]
14. Russell, A.D. Lethal Effects of Heat on Bacterial Physiology and Structure. *Sci. Prog.* **2003**, *86*, 115–137. [CrossRef] [PubMed]
15. Cebrián, G.; Condón, S.; Mañas, P. Physiology of the Inactivation of Vegetative Bacteria by Thermal Treatments: Mode of Action, Influence of Environmental Factors and Inactivation Kinetics. *Foods* **2017**, *6*, 107. [CrossRef] [PubMed]
16. Pohl, H.A.; Hawk, I. Separation of Living and Dead Cells by Dielectrophoresis. *Science* **1966**, *152*, 647–649. [CrossRef] [PubMed]
17. Markx, G.H.; Talary, M.S.; Pethig, R. Separation of viable and non-viable yeast using dielectrophoresis. *J. Biotechnol.* **1994**, *32*, 29–37. [CrossRef]
18. Markx, G.H.; Pethig, R. Dielectrophoretic separation of cells: Continuous separation. *Biotechnol. Bioeng.* **1995**, *45*, 337–343. [CrossRef]
19. Lapizco-Encinas, B.H.; Simmons, B.A.; Cummings, E.B.; Fintschenko, Y. Dielectrophoretic Concentration and Separation of Live and Dead Bacteria in an Array of Insulators. *Anal. Chem.* **2004**, *76*, 1571–1579. [CrossRef]
20. Ho, B.D.; Beech, J.P.; Tegenfeldt, J.O. Charge-Based Separation of Micro- and Nanoparticles. *Micromachines* **2020**, *11*, 1014. [CrossRef]
21. Karabacak, N.M.; Spuhler, P.S.; Fachin, F.; Lim, E.J.; Pai, V.; Ozkumur, E.; Martel, J.M.; Kojic, N.; Smith, C.L.; Chen, P.-I.; et al. Microfluidic, marker-free isolation of circulating tumor cells from blood samples. *Nat. Protoc.* **2014**, *9*, 694–710. [CrossRef]

vessel due to the instant release of fluid and microspheres at high pressure. To this end, microsphere blockages in catheters must be resolved for proper embolism practices.

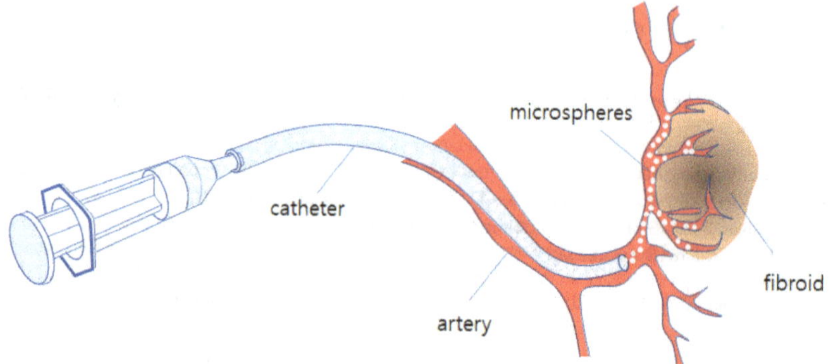

Figure 1. Schematic of fibroid embolization of microspheres with catheter.

There may be two possible mechanisms on the blockage of microspheres flowing with a carrier liquid in the catheter [11]. One is the geometrical arching of the microspheres, and the other is the electrical deposition on the wall surface. The arching is a clogging phenomenon due to the steric effect in which the microspheres form an arch shape in the catheter channel [12–14]. Electrical deposition occurs when one microsphere attaches onto the surface of the channel wall due to electrical attractive force. Then, other microspheres are successively attached to the wall, and the attached microspheres form a lump gradually. The lump also gradually increases to block the catheter [15–17].

Many studies have been done on blockage, but few studies have introduced methods to solve the blockages. There are two representative methods to solve it. One is to remove the blockage of the microspheres with momentary motions by exerting oscillating waves [18–21]. This method constantly consumes power to use an amplifier and requires connecting the channel to a separate electrical device. The other method is to reduce the deposition of microspheres by inducing a repulsive electrical force chemically on the surface [22]. This requires an additional and sophisticated chemical treatment on the surface of the catheter.

This study proposes a novel method to prevent clogging by microspheres in a catheter by using convex air bubbles attached to the inner surface of the catheter. Convex air bubbles have a slip effect on their bubble surface and a centrifugal effect on the microspheres due to their small radius of curvature. In a fluid flow, the velocity is zero on the stationary wall, which is called the no-slip condition. However, the measurement of the micro-PIV(particle image velocimetry) over the surface of a convex air bubble shows that there exists a velocity with a certain value, which indicates the occurrence of the slip on it [23]. The slippery surface of the convex air bubble can reduce the friction, which has a key role in forming the blockage arching of microspheres in the catheter channel.

In addition, when the microsphere travels along the convex air bubble, centrifugal acceleration normal to the surface is obtained due to the microsphere's small radius of curvature and is exerted on the microsphere. Even though it is a very low Reynolds flow, it makes the microsphere escape from the convex surface. Thus, the distance is increased between the microsphere and the channel wall. This centrifugal effect can reduce the attachment of microspheres so that it can prevent their deposition. As a result, both the friction reduction on the surface by the slip effect and the increased distance away from the catheter wall can effectively prevent the blockages caused by microsphere clogging.

To determine these effects by the convex air bubbles, straight catheter channels with microcavities were fabricated by the MEMS fabrication process. Additionally, a straight channel without cavities was fabricated as a reference, and the clogging of the microspheres was compared. The formation of the convex air bubbles in the cavities was visualized, and the two effects were experimentally examined and theoretically explained.

2. Fabrication of the Catheter Channel with Cavities

The term micro generally means a size from 1 µm to 1 mm. As the characteristic size becomes smaller, the ratio of the surface area to volume increases. Thus, the convex air bubbles attached in the microcavities hardly detach from the surface because the buoyancy and the shear force as a detaching force are much smaller than the surface tension as an attaching force [24].

The embolization catheter is generally cylindrical. The hydrophobic cylindrical tubes are manufactured by an extrusion process. However, in order to form the air bubble for the introduction of a liquid, the tube must have cavities on the inner surface. This makes the manufacturing process very complicated and challenging. In addition, it is very difficult to visualize the movement of the microspheres in the 3-dimensional cylindrical tube. So, for the experimental evaluation, the catheter channel with microcavities was fabricated precisely by using MEMS fabrication technology. Then, the channel surface was modified hydrophobically. Convex air bubbles naturally form in the hydrophobic cavities as soon as water is introduced.

To examine the effect of the convex air bubbles on the prevention of catheter blockage by microspheres, straight catheter channels were fabricated with and without micro-cavities. The length of the catheter channel was 40 mm. Because it is hard to specify the exact position where the blockage occurs, two pressure ports were connected to the catheter channel to detect the sudden pressure build-up by the blockage, and the distance between them was 32 mm. Microspheres with a diameter of 100 µm have been used clinically in embolization [25]. Thus, a channel height of 150 µm was designed by considering their size. Therefore, only one microsphere can be placed in the depth direction. Two-dimensional blockage arching can be situated. Additionally, the width of the catheter channel was varied and included 200, 300, and 400 µm, by considering the number of microspheres involved in the arching. Moreover, by considering the slip influence distance of a single bubble as 209 µm, which was reported in a previous work, two different intervals of 100 and 500 µm were used, indicating the distance between the cavities [23].

The catheter channel connected with the cavities and the pressure ports was fabricated by soft lithography. First, the mold pattern was prepared with photolithography. To make the mold pattern of the catheter channel, the adhesion promoter of HDMS was spin-coated on a 4-inch silicon wafer at 4000 rpm for 30 s. It was baked on a hot plate at 150 °C for 2 min. SU-8 100 photo-resist was coated at 500 rpm for 10 s and 1750 rpm for 30 s. It was cured on a hot plate at 65 °C for 10 min and at 95 °C for 30 min. Then, it was exposed to UV for 37 s for photolithography. Again, the exposed silicon wafer was placed on the hot plate and baked at 65 °C for 1 min and at 95 °C for 12 min [26]. The baked silicon wafer was cooled down and developed in a SU-8 developer until the catheter channel patterns were successfully obtained. Then, it was washed with isopropyl alcohol and de-ionized water several times and dried by heating it at 150 °C for 3 min.

Secondly, the PDMS(polydimethylsiloxane) microchannel catheter was fabricated by molding the PDMS. The inserted eyelets of the silicone tube were aligned and bonded to the inlet and outlet of the patterned SU-8 catheter channel and pressure ports, respectively. Then, the PDMS mixed with a pre-polymer and curing agent (Sylgard 184, Dow Corning, MI, USA) with a ratio of 10:1 was poured onto the patterned silicon wafer at a thickness of 5 mm. After removing the air bubbles remaining inside the PDMS in a vacuum oven, it was cured at 60 °C for 2 h to minimize the PDMS shrinkage.

Finally, each device was cut and exposed to oxygen plasma for 2 min with slide glass. This process makes the surface hydrophilic. Then, the mold PDMS and the slide glass were bonded naturally. The fabricated catheter channel was inspected with a microscope and SEM images. The inlet and outlet at both ends were successfully connected to the microchannel catheter as aligned. The height and width were measured to be 160 µm with a standard deviation of 3.68 µm and 426 µm with a standard of 3.71 µm, respectively. The cavity size was 106 µm with a standard deviation of 2.38 µm.

3. Clogging Prevention Experiment

In general, PVA (polyvinyl alcohol) or gelatin spheres are mostly used for the embolus microspheres. In this experiment, glass microspheres with a specific gravity of 2.5 were used due to the difficulty in the microscopic visualization of PVA spheres. The glass microspheres settled to the channel bottom and their movement was examined on the microscope. The size of the glass microspheres was from 90 to 106 µm and water were prepared as a carrier fluid. A high-speed camera (VEO 640L, Phantom, NJ, USA) was set up in a vertical microscope (IX 71, Olympus, Tokyo Japan) to visualize the movement of the microspheres and their blockage in the catheter microchannel. A syringe pump (PHD 2000, Harvard apparatus, Holliston, MA, USA) was connected to the inlet and introduced the microsphere-suspended water. The microspheres were injected randomly. It was hard to visualize the exact position of the occurrence of the blockage in the 40 mm-long channel in real-time due to the limitation of the microscope lens. Thus, to indicate the occurrence of the blockage, the differential pressure was measured between the two pressure ports. Moreover, the convex air bubbles were formed in the cavities by introducing pure de-ionized water in advance.

3.1. Slip Effect of the Convex Air Bubble

It was observed that the glass microspheres easily settled to the bottom of the catheter. Thus, in order to carry the microspheres stably in water, the Reynolds number of the flow was examined. By considering the hydraulic diameter of the catheter microchannel, Reynolds numbers ranging from 3.6 to 72.2 were tested, and it was found that microspheres could move stably with a Reynolds number over 20 [27].

The two straight microchannels with and without the air bubbles were compared. The channel width was 200 µm. The inlet flow rate was 257.1 µL/min, corresponding to an Re of 20. In the straight microchannel without the air bubbles, blockage occurred. It was hard to control the exact condition and position of the blockage occurrence. Only after the blockage occurred were its characteristics measured. Figure 2a shows a picture taken after the blockage occurred in the catheter channel without the air bubbles. Its angle was obtained from the line connecting the centers of the two microspheres and the channel length. The blockage arching angle was obtained at an angle of about 70° from the image analysis, similar to one theoretically obtained. However, it does not mean that the measured blockage angle of 70° is a threshold or minimal angle.

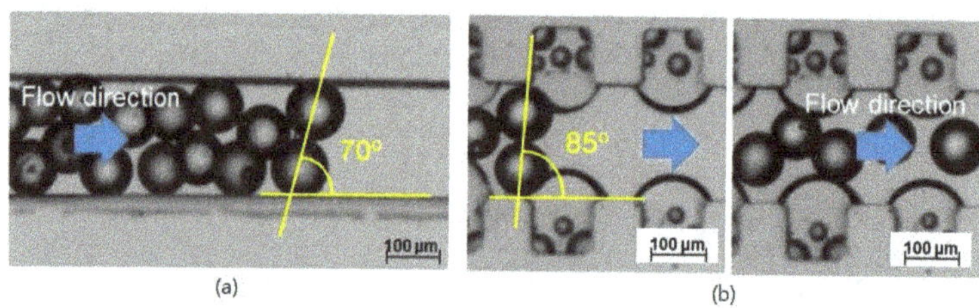

Figure 2. Prevention of microsphere blockage in the catheter channel by convex air bubbles; (a) Occurrence of the blockage arching at an arching angle of 70° in the straight catheter channel without the convex air bubbles, (b) Free movement of the microspheres at a geometrical arching angle of over 85° in the straight catheter channel with the convex air bubbles.

In the straight channel attached with the convex air bubbles formed in the microcavities, the microspheres slide freely without any blockage from the arching despite a geometrical arching angle of 85° due to the slippery and deformable surface of the convex air bubbles shown in Figure 2b.

Depending on the size of the catheter channel and the microspheres, the number of microspheres involved in the blockage arching can be determined. Figure 3 shows the 2-dimensional blockage arching involved with three microspheres in the straight catheter channel. The plane is positioned in the center where z equals 0 in height. The force diagram includes all the forces acting on the microsphere positioned in the middle and two microspheres pushed toward the wall. Three important forces are involved for the blockage arching. They are the drag force caused by the fluid flow, the friction force between the wall surface and the microsphere, and the reaction force between the two microspheres as a response to action and reaction.

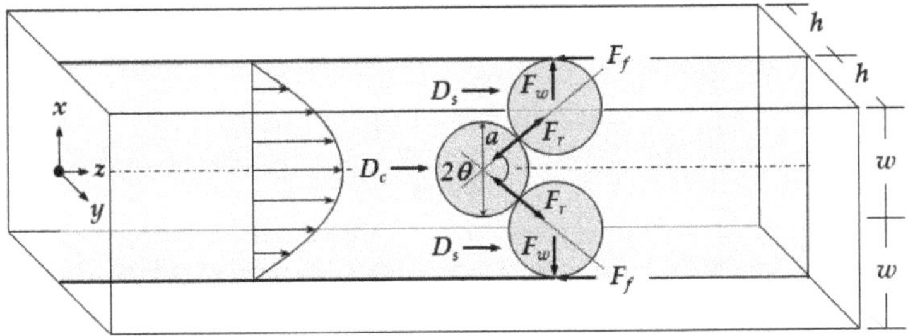

Figure 3. Diagram of the forces acting on the microspheres when the blockage occurs in catheter by arching.

The drag force was analyzed with the assumptions of a steady and incompressible flow. Additionally, a 2-dimensional fully developed velocity profile was considered along the catheter channel for the sake of convenience, even though the velocity profile is parabolic 3-dimensionally in the rectangular catheter channel with the sizes of the channel height of 160 μm, width of 426 μm, and length of 40 mm. In addition, the velocity profile could be affected by the existence of the microspheres. Due to a very low Reynolds number, a strong laminar flow is formed, and the microspheres move along the streamlines since the movement of the spheres is governed mainly by fluidic force. Therefore, in the calculation their effect on the velocity profile was also neglected.

The 2-dimensional velocity distribution was obtained on the plane where $y = 0$ in the z-axis direction as Equation (1) [28,29]. The inlet flow rate, Q_{in}, was calculated by integrating the velocity distribution over the width as Equation (2).

$$u_z(x, y = 0) = \frac{16\alpha^2 h^2}{\mu \pi^4}\left(\frac{\Delta p}{l}\right)\sum_{n=1}^{\infty}\sum_{m=1}^{\infty}\frac{(1 - \cos m\pi)(1 - \cos n\pi)}{mn(m^2 + \alpha^2 n^2)}\sin\frac{n\pi}{2}\sin\frac{m\pi}{2w}(x + w) \tag{1}$$

$$Q_{in} = \int_{-w}^{w}u_z(x)dx = \frac{32\alpha^3 h^3}{\mu \pi^5}\left(\frac{\Delta p}{l}\right)\sum_{n=1}^{\infty}\sum_{m=1}^{\infty}\frac{(1 - \cos m\pi)^2(1 - \cos n\pi)}{m^2 n(m^2 + \alpha^2 n^2)}\sin\frac{n\pi}{2} \tag{2}$$

where α is the aspect ratio of the catheter channel, μ is the viscosity of the fluid, and Δp is the pressure drop according to channel length, l. Then, the 2-dimensional velocity distribution can be expressed as Equation (3) by substituting Equation (2) into Equation (1).

$$u_z(x) = Q_{in}\frac{\pi}{2w}\frac{\sum_{n=1}^{\infty}\sum_{m=1}^{\infty}\frac{(1-\cos m\pi)(1-\cos n\pi)}{mn(m^2+\alpha^2 n^2)}\sin\frac{n\pi}{2}\sin\frac{m\pi}{2w}(x+w)}{\sum_{n=1}^{\infty}\sum_{m=1}^{\infty}\frac{(1-\cos m\pi)^2(1-\cos n\pi)}{m^2 n(m^2+\alpha^2 n^2)}\sin\frac{n\pi}{2}} \tag{3}$$

Assume that two microspheres attached on the wall surface are placed symmetrically, and the flow rates on them, Q_l and Q_u, are obtained by subtracting the flow rate on the middle microsphere, Q_c, from the inlet flow rate as Equation (4).

$$Q_l = Q_u = \frac{Q_{in} - Q_c}{2}$$

$$= \frac{Q_{in}}{2}\left[1 - \frac{\sum_{n=1}^{\infty}\sum_{m=1}^{\infty}\frac{(1-\cos m\pi)(1-\cos n\pi)}{m^2 n(m^2 + a^2 n^2)}\sin\frac{n\pi}{2}\left[\cos\frac{m\pi}{2w}\left(w - \frac{a}{2}\right) - \cos\frac{m\pi}{2w}\left(w + \frac{a}{2}\right)\right]}{\sum_{n=1}^{\infty}\sum_{m=1}^{\infty}\frac{(1-\cos m\pi)^2(1-\cos n\pi)}{m^2 n(m^2 + a^2 n^2)}\sin\frac{n\pi}{2}}\right] \tag{4}$$

The average velocity is then obtained by dividing the flow rate entering into each microsphere with the confronted projection area. The average velocity on the center microsphere, $\overline{V}_c$, is calculated in Equation (5).

$$\overline{V}_c = \frac{Q_c}{a}$$

$$= \frac{\sum_{n=1}^{\infty}\sum_{m=1}^{\infty}\frac{(1-\cos m\pi)(1-\cos n\pi)}{m^2 n(m^2 + a^2 n^2)}\sin\frac{n\pi}{2}\left[\cos\frac{m\pi}{2w}\left(w - \frac{a}{2}\right) - \cos\frac{m\pi}{2w}\left(w + \frac{a}{2}\right)\right]}{a\sum_{n=1}^{\infty}\sum_{m=1}^{\infty}\frac{(1-\cos m\pi)^2(1-\cos n\pi)}{m^2 n(m^2 + a^2 n^2)}\sin\frac{n\pi}{2}} \tag{5}$$

Because the open area to two microspheres is $(2w - a)$, the position of their average velocity does not coincide with the geometric center of the microspheres. The position of the average velocity of the microsphere on the wall is obtained by dividing the flow rate with the open area, as shown in Equation (6).

$$\overline{V}_l = \overline{V}_u = \frac{Q_l}{\frac{2w-a}{2}} = \frac{Q_{in}}{2w-a}\left[1 - \frac{\sum_{n=1}^{\infty}\sum_{m=1}^{\infty}\frac{(1-\cos m\pi)(1-\cos n\pi)}{m^2 n(m^2 + a^2 n^2)}\sin\frac{n\pi}{2}\left[\cos\frac{m\pi}{2w}\left(w - \frac{a}{2}\right) - \cos\frac{m\pi}{2w}\left(w + \frac{a}{2}\right)\right]}{\sum_{n=1}^{\infty}\sum_{m=1}^{\infty}\frac{(1-\cos m\pi)^2(1-\cos n\pi)}{m^2 n(m^2 + a^2 n^2)}\sin\frac{n\pi}{2}}\right] \tag{6}$$

The drag force exerted on the microsphere is known as the Stokes force in a low Reynolds flow. So, each drag force of D_c, D_l, and D_u is given by Equations (7) and (8), where μ is the dynamic viscosity of the fluid.

$$D_c = 6\pi\mu\left(\frac{a}{2}\right)\overline{V}_c \tag{7}$$

$$D_l = D_u = 6\pi\mu\left(\frac{a}{2}\right)\overline{V}_l = 6\pi\mu\left(\frac{a}{2}\right)\overline{V}_u \tag{8}$$

The center of the drag force is assumed to be identical to the center of the mean velocity. By considering the drag force and the center position, the blockage arching angle, θ, can be obtained from Equation (9).

$$\theta = sin^{-1}\frac{2w - a}{2a} \tag{9}$$

The action-reaction force of F_r can be obtained from the drag force, D_c, acting on the microsphere positioned in the middle and the blockage arching angle, θ.

$$F_r = \frac{D_c}{2cos\theta} \tag{10}$$

Finally, the friction force of F_f is obtained as follows.

$$F_r = \mu_f F_N = \mu_f F_r sin\theta \tag{11}$$

where μ_f is the coefficient of the static friction. In general, the static friction is larger than the dynamic friction. The three important forces to form the blockage arching of microspheres in the catheter channel were calculated. For $\mu_f = 0.7$, $a = 100$–145 μm, $w = 150$ μm, $\alpha = 2.5$, and $\mu = 1.12 \times 10^{-3}$ N/m·s, the blockage arching angle was analyzed. If the summation of drag forces is smaller than the friction force, it is possible for a blockage to occur. Figure 4 shows the blockage arching angle, and it was determined to be over 76°, approximately. In addition, the blockage angle depends on the material of the microspheres because the coefficient of the static friction differs by material.

From the experimental results, large convex air bubbles and a small interval between the cavities were effective at preventing microsphere blockage. Thus, it can be concluded that it will be advantageous to design a catheter channel with large cavities and a small interval for the prevention of microsphere blockage in embolization. For further study, the occurrence of blockages from the bending of the catheter, which often happens during embolization, will be examined by considering the radius of the curvature of the catheter.

Supplementary Materials: The following are available online at http://www.mdpi.com/2072-666X/11/12/1040/s1, Video S1: Microspheres in the catheter channel.

Author Contributions: D.H.P. has contributed to the design of size and shape and experiment of this work. Y.J.J. has contributed to the theoretical calculation and experiment to visualize the movement of microspheres. S.J.K.S.J.K. and Y.D.K. helped to fabricate the microchannel catheter. J.S.G. as corresponding author managed whole work. All authors have read and agreed to the published version of the manuscript.

Funding: This research was financially supported by the National Research Foundation of Korea (NRF) grant funded by the Korea government (MOE) (No. NRF-2017R1A2B2006264).

Conflicts of Interest: The authors declare no conflict of interest.

References

1. Ryu, E.H. Nano structure-based drug delivery technology. *J. KSME* **2012**, *52*, 43–47.

2. Damiati, S.; Kompella, U.B.; Damiati, S.A.; Kodzius, R. Microfluidic Devices for Drug Delivery Systems and Drug Screening. *Genes* **2018**, *9*, 103. [CrossRef]

3. Greish, K. Enhanced Permeability and Retention (ERP) Effect for Anticancer Nanomedicine Drug Targeting. *Methods. Mol. Biol.* **2010**, *624*, 25–37. [CrossRef]

4. Jones, C.N.; Hoang, A.N.; Dimisko, L.; Hamza, B.; Martel, J.; Irimia, D. Microfluidic Platform for Measuring Neutrophil Chemotaxis from Unprocessed Whole Blood. *J. Vis. Exp.* **2014**, *88*. [CrossRef] [PubMed]

5. Bernacka-Wojcik, I.; Lopes, P.; Vaz, A.C.; Veigas, B.; Wojcik, P.J.; Simões, P.; Barata, D.; Fortunato, E.; Baptista, P.V.; Águas, H.; et al. Bio-microfluidic Platform for Gold Nanoprobe Based DNA Detection-application to Mycobacterium tuberculosis. *Biosens. Bioelectron.* **2013**, *48*, 87–93. [CrossRef] [PubMed]

6. Peng, Z.; Young, B.; Baird, A.E.; Soper, S.A. Single-pair Fluorescence Resonance Energy Transfer Analysis of mRNA Transcripts for Highly Sensitive Gene Expression Profiling in Near Real Time. *Anal. Chem.* **2013**, *85*, 7851–7858. [CrossRef] [PubMed]

7. Li, X.; Valadez, A.; Zuo, P.; Nie, Z. Microfluidic 3D Cell Culture: Potential Application for Tissue-based Bioassays. *Bioanalysis* **2012**, *4*, 1509–1525. [CrossRef]

8. Park, D.H.; Jeon, H.J.; Kim, M.J.; Nguyen, X.D.; Morten, K.; Go, J.S. Development of a Microfluidic Perfusion 3D Cell Culture System. *J. Micromech. Microeng.* **2018**, *28*, 1–9. [CrossRef]

9. Harrigan, M.R.; Deveikis, J.P. *Handbook of Cerebrovascular Disease and Neurointerventional Technique*, 1st ed.; Harrigana, M.R., Deveikis, J.P., Eds.; Humana Press: New York, NY, USA, 2009.

10. Lewis, A.L.; Adams, C.; Busby, W.; Jones, S.A.; Wolfenden, L.C.; Leppard, S.W.; Palmer, R.R.; Small, S. Comparative in vitro evaluation of microspherical embolisation agents. *J. Mater. Sci. Mater. Med.* **2006**, *17*, 1193–1204. [CrossRef]

11. Dressaire, E.; Sauret, A. Clogging of microfluidic systems. *Soft Matter* **2017**, *13*, 37–48. [CrossRef]

12. Sharp, K.V.; Adrian, R.J. On flow-blocking microsphere structures in microtubes. *Microfluid. Nanofluid.* **2005**, *1*, 376–380. [CrossRef]

13. Agbangla, G.C.; Bacchin, P.; Climent, E. Collective Dynamics of Flowing Colloids During Pore Clogging. *Soft Matter* **2014**, *10*, 6303–6315. [CrossRef] [PubMed]

14. Agbangla, G.C.; Climent, E.; Bacchin, P. Experimental Investigation of Pore Clogging by Microparticles: Evidence for a Critical Flux Density of Particle Yielding Arches and Deposits. *Sep. Purif. Technol.* **2012**, *101*, 42–48. [CrossRef]

15. Dersoir, B. Clogging in micro-channels: From colloidal microsphere to clog. *HAL* **2015**, *1*, 376–380.

16. Dersoir, B.; de Saint Vincent, M.R.; Abkarian, M.; Tabuteau, H. Clogging of a single pore by colloidal microspheres. *Microfluid. Nanofluid.* **2015**, *19*, 953–961. [CrossRef]

17. Dersoira, B.; Schofieldb, A.B.; de Saint Vincenta, M.R.; Tabuteaua, H. Dynamics of pore fouling by colloidal microspheres at the microsphere level. *J. Membr. Sci.* **2019**, *573*, 411–424. [CrossRef]

18. Sriphutkiat, Y.; Zhou, Y. Particle Accumulation in a Microchannel and Its Reduction by a Standing Surface Acoustic Wave (SSAW). *Sensors* **2017**, *17*, 106. [CrossRef]

19. Ding, X.; Lin, S.S.; Kiraly, B.; Yue, H.; Li, S.; Chiang, I.; Shi, J.; Benkovic, S.J.; Huang, T.J. On-chip Manipulation of Single Microparticles, Cells, and Organisms Using Surface Acoustic Waves. *Proc. Natl. Acad. Sci. USA* **2012**, *109*, 11105–11109. [CrossRef]

20. Destgeer, G.; Lee, K.H.; Jung, J.H.; Alazzam, A.; Sung, H.J. Continuous Separation of Particles in a PDMS Microfluidic Channel via Travelling Surface Acoustic Waves (TSAW). *Lab Chip.* **2013**, *13*, 4210–4216. [CrossRef]

21. Ding, X.; Peng, Z.; Lin, S.-C.S.; Geri, M.; Li, S.; Li, P.; Chen, Y.; Dao, M.; Suresh, S.; Huang, T.J. Cell separation using tilted-angle standing surface acoustic waves. *Proc. Natl. Acad. Sci. USA* **2014**, *111*, 12992–12997. [CrossRef]

22. Sardella, E.; Detomaso, L.; Gristina, R.; Senesi, G.S.; Agheli, H.; Sutherland, D.S.; d'Agostino, R.; Favia, P. Nano-Structured Cell-Adhesive and Cell-Repulsive Plasma-Deposited Coating: Chemical and Topographical Effects on Keratinocyte Adhesion. *Plasma Processes Polym.* **2008**, *5*, 540–551. [CrossRef]

23. Kwon, B.H. Experimental study on the reduction of skin frictional drag in pipe flow by using convex air bubbles. *Exp. Fluids* **2014**, *55*, 1722. [CrossRef]

24. Tesař, V. What can be done with microbubbles generated by a fluidic oscillator? (survey). *EPJ Web Conf.* **2017**, *143*, 2129. [CrossRef]

25. Doucet, J.; Kiri, L.; O'Connell, K.; Kehoe, S.; Lewandowski, R.J.; Liu, D.; Abraham, R.J.; Boyd, D. Advances in Degradable Embolic Microspheres: A State of the Art Review. *J. Funct. Biomater.* **2018**, *9*, 14. [CrossRef] [PubMed]

26. Sensu, Y.; Sekiguchi, A.; Kono, Y. Modeling of Cross-linking Reactions for Negative-type Thick film Resists. *J. Photopolym. Sci. Technol.* **2006**, *19*, 51–56. [CrossRef]

27. Caine, M.; Carugo, D.; Zhang, X.; Hill, M.; Dreher, M.R.; Lewis, A.L. Review of the Development of Methods for Characterization of Microspheres for Use in Embolotherapy: Translating Bench to Cathlab. *Adv. Health Mater.* **2017**, *6*. [CrossRef] [PubMed]

28. Kashaninejad, N. A New Form of Velocity Distribution in Rectangular Microchannels with Finite Aspect Ratios. *Preprints* **2019**. [CrossRef]

29. Zheng, X.; Silber-Li, Z. Measurement of velocity profiles in a rectangular microchannel with aspect ratio α = 0.35. *Exp. Fluids* **2008**, *44*, 951–959. [CrossRef]

30. Squires, T.M.; Quake, S.R. Microfluidics: Fluid physics at the nanoliter scale. *Rev. Mod. Phys.* **2005**, *77*, 977–1026. [CrossRef]

Publisher's Note: MDPI stays neutral with regard to jurisdictional claims in published maps and institutional affiliations.

 *micromachines*

Article

Charge-Based Separation of Micro- and Nanoparticles

Bao D. Ho, **Jason P. Beech** and **Jonas O. Tegenfeldt** *

Division of Solid State Physics and NanoLund, Physics Department, Lund University, P.O. Box 118, 22100 Lund, Sweden; bao.hodang@gmail.com (B.D.H.); jason.beech@ftf.lth.se (J.P.B.)
* Correspondence: jonas.tegenfeldt@ftf.lth.se; Tel.: +46-46-222-8063

Received: 16 October 2020; Accepted: 14 November 2020; Published: 18 November 2020

Abstract: Deterministic Lateral Displacement (DLD) is a label-free particle sorting method that separates by size continuously and with high resolution. By combining DLD with electric fields (eDLD), we show separation of a variety of nano and micro-sized particles primarily by their zeta potential. Zeta potential is an indicator of electrokinetic charge—the charge corresponding to the electric field at the shear plane—an important property of micro- and nanoparticles in colloidal or separation science. We also demonstrate proof of principle of separation of nanoscale liposomes of different lipid compositions, with strong relevance for biomedicine. We perform careful characterization of relevant experimental conditions necessary to obtain adequate sorting of different particle types. By choosing a combination of frequency and amplitude, sorting can be made sensitive to the particle subgroup of interest. The enhanced displacement effect due to electrokinetics is found to be significant at low frequency and for particles with high zeta potential. The effect appears to scale with the square of the voltage, suggesting that it is associated with either non-linear electrokinetics or dielectrophoresis (DEP). However, since we observe large changes in separation behavior over the frequency range at which DEP forces are expected to remain constant, DEP can be ruled out.

Keywords: electrokinetic deterministic lateral displacement; charge-based separation

1. Introduction

Deterministic Lateral Displacement (DLD) is a powerful size-based particle sorting method that has multiple advantages: It is label-free, it has high resolution, and it allows for continuous operation [1]. The method has been applied to sort various types of cells or bio-particles from blood: white blood cells [2–6], circulating tumor cells [7–12], trypanosomes [13,14], nucleated red blood cells [15], microvesicles [16,17], and PC3 prostate cancer cells [18]. In addition, DLD has found applications in separation or enrichment of different types of mammalian cells [19–23], fungal spores [24], droplets [25,26], bacteria [27], DNA [1,28], and exosomes [29]. Extensive reviews of DLD can be found elsewhere [30,31]. Here we give a brief description of the method.

A schematic of a DLD device is shown in Figure 1a. The main feature is an array of micro-pillars that is tilted at an angle with respect to the flow carrying the particles. A heterogeneous sample containing small and large particles is loaded into the sample reservoir on the top left of the schematic in Figure 1a and driven through the pillar array. The DLD mechanism is fundamentally based on whether or not particles are able to follow the carrier fluid as they moves through the array. At low Reynolds numbers it is primarily steric interactions between the particles and the posts that prevent this but under certain conditions other interactions play a role such as charge interactions when the carrier fluid has low ionic strength [32] or inertia and wall lift forces when Reynolds numbers are high [33]. These additional interactions can be carefully chosen to be sensitive to relevant particle parameters and we will show here how externally applied AC fields make it possible to sort particles by their zeta potential in a tunable manner.

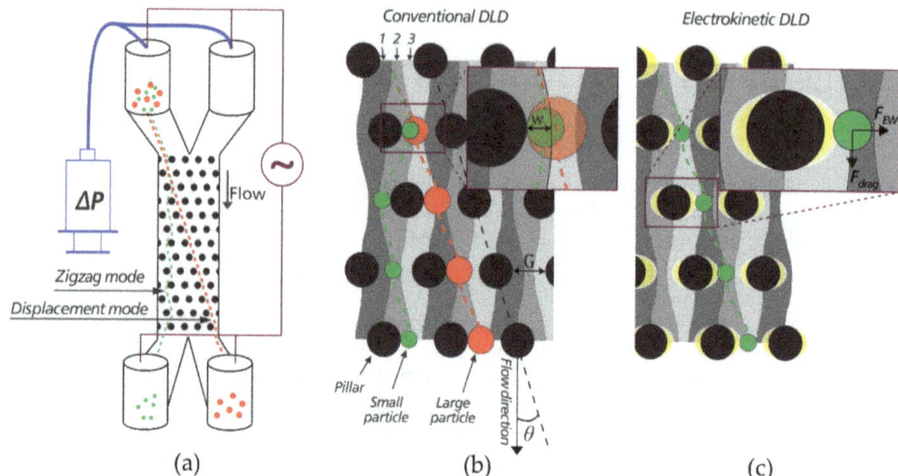

Figure 1. Working principle of conventional Deterministic Lateral Displacement (DLD) and DLD with electric fields (eDLD). (**a**) Diagram of a typical DLD device. The device is driven by a pressure difference between inlets and outlets and is capable of sorting a mixture of particles by size. Electrokinetic effects are added by applying a voltage along the device. In this case, the average field is parallel to the flow. (**b**) Mechanism of conventional DLD. Separation is a result of consecutive migration of large particles across different flow streams (here numbered 1, 2, 3) due to steric interactions with the pillars. (**c**) In eDLD, the particle trajectories are modified by electrokinetic wall force fields around the pillars.

Unlike sieves, which also hinder particles from following the flow, steric and other interactions lead to changes in trajectories only and DLD devices can therefore be considered as non-clogging sieves. The DLD mechanism causes different-sized particles to follow different trajectories so that they can be collected in separate reservoirs at the end of the device.

The mechanism of conventional DLD (steric interactions only) is illustrated in Figure 1b. The pillar array repeats itself after N rows where N is called the period of the DLD. Due to the specific arrangement of the pillar array, the fluid flow is split into N different streams between any two neighboring pillars in each row. For a device with equal lateral and longitudinal inter-post distance, the corresponding tilting angle of the array is given by $tan(\theta) = 1/N$. For the sake of convenience and to avoid anomalous modes [34], the design is often such that N is an integer. As an example, in Figure 1b, the period is $N = 3$, corresponding to a tilt of $arctan(1/3) = 18.4°$.

The separate trajectories of one small particle (green) and one large particle (red) are illustrated in Figure 1b. The radius of the green particle is smaller than the stream width w depicted in Figure 1b, and as a result, the particle's center of mass remains in stream #2 throughout the array. This mode of transport is termed the zigzag mode due to the shape of the trajectories. In contrast, the red particle whose radius is larger than w bumps into the pillar and is displaced from stream #2 to stream #3. This stream-switching occurs every time the particle encounters a new row of pillars, forcing the particle to travel along the tilting angle of the array. This mode is termed the *displacement mode* and we will refer to the angle θ henceforth as the *displacement angle*. At the end of one period (N rows) of the pillar array, particles in zigzag mode have a maximum lateral displacement of $G + D$, where G is the gap between posts and D is the post diameter, while particles in displacement mode are shifted laterally by a distance $N \times (G + D)$. Using the angular formulation instead, the displacement of the smaller particles is ≈ 0 and that of the large particles is $l \times tan(\theta)$, where l is the length of one period (N rows) of the pillar array. The *critical diameter* D_C, which is the diameter at which the transition between zigzagging and displacing occurs, is predetermined by the geometry of the array, and can be estimated by:

$$D_C = 1.4GN^{-0.48} \tag{1}$$

Equation (1), an empirical formula reported by Davis [35], is used throughout this paper to estimate the critical diameters of our DLD devices.

Although the behaviors of particles in DLD devices are primarily dependent on their size, size is not always a well-defined property of a particle. Instead, we use the term effective size to relate the particle's trajectory to the trajectory of a perfectly spherical particle moving through a DLD device driven solely by a pressure driven flow. In other words, an arbitrary particle under arbitrary experimental conditions is said to have an effective size D_{eff} if it moves along the same trajectory as a perfectly spherical particle of size $D_{particle} = D_{eff}$ experiencing only the pressure driven flow and steric interactions in the DLD array. This is useful to describe, for example, non-spherical, deformable, or charged particles that have effective sizes that depend on the detailed properties of the particles along with the exact experimental conditions being used. Previous approaches have targeted one or more of these properties to separate particles in DLD based on, for example, shape [13,14,36,37], length [27], and deformability [6,36,38]. Surface charge has also been employed in the context of DLD [32,39], but in general in a different manner than in our current work. Zeming et al. [32] modulated the effective size by tuning the electrostatic force between pillars and nanoparticles, which was achieved by adjusting the salt concentration of the running medium. The method works well for particles below 1 μm in low-salt medium ($\leq$1 mM NaCl). In contrast, our approach, with the integration of an electric field, can be used for micro as well as nano particles, and operates at higher salt concentrations that can be relevant for biological samples.

The interaction between particles and applied electric fields is complicated. A full description must take into account the behavior of the solvent (water) and the dissolved ions near the particle surface. The zeta potential of a particle is simply the electrostatic potential at the shear plane due to the charges on the particle. The zeta potential depends on the density of charged groups on the surface of the particle, the distance between the particle surface and the shear plane, and the ionic strength of the suspending medium. Because the zeta potential is both easy to measure (taking several minutes with a Zetasizer NanoZS instrument (see Section 2.1) and captures the behavior of the effective charge of the particle as the ionic strength of the buffer changes, we will use "zeta potential" throughout and not "charge", but the reader should keep in mind that they are related.

The use of electric fields for particle manipulation in a liquid medium encompasses multiple mechanisms, among which electrophoresis and dielectrophoresis are the most notable. Electrophoresis (EP) refers to the movement of charged particles in an electric field, which can be uniform or non-uniform [40–43]. It should be noted that EP of charged particles is in fact "force-free": the particle motion is driven by the electroosmotic slip on the particle surface. The applications of electrophoresis to manipulate and separate particles and molecules are countless, and it has become a gold standard for separation of proteins and nucleic acids. Dielectrophoresis (DEP) refers to the movement of polarizable particles (charged or uncharged) in a polarizable medium under a non-uniform electric field [44–46]. Although having a shorter history than electrophoresis, DEP has also found numerous applications in enrichment, isolation, or separation of bio-particles, including yeast cells [47–50], human cells [51], microalgae [52], bacteria [53–55], viruses [56], DNA [57,58], proteins [59], and nanocrystals [60], to name but a few. Dielectrophoretic techniques can be categorized into Microelectrode-based DEP (eDEP) or Insulator-based DEP (iDEP) [61]. As their names suggest, these two branches are distinguished by the materials the particles are in contact with: conducting electrodes (eDEP) or insulating obstacles (iDEP). For eDEP systems, in the early days, the electrodes were pin-and-plate macroelectrodes [47] but using microfabrication, electrodes have been miniaturized to achieve higher field strengths. Various configurations of electrodes have been reported in the literature, with castellated [48,49,56,62,63], quadrupole polynomial [56,64], zipper [65], or 3D-well [66] designs. In iDEP systems, efforts are invested in designing innovative insulating structures inside microfluidic channels to achieve the desired field gradients for DEP. These structures include constrictions [51,59,60,67–69], ridges [70,71], curves [50,72], and obstacle arrays [53,57,73–75]. A thorough review of iDEP can be found elsewhere [61].

In terms of configuration, the electrokinetic DLD devices presented in this paper are relatively similar to an iDEP system with an insulating pillar array. The key difference is, however, that our devices have pillar arrays tilted at an angle to the flow to achieve the DLD function. Furthermore, we are interested in not only DEP, but also other electrokinetic phenomena that enhance particle displacement. In general, any electrokinetic force fields that are specific to some particle property and that change the probability that particles switch streams at every row of pillars (altering the cumulative lateral displacement of the particles), can be exploited for tunable particle separations (Figure 1c). The nature of different electrokinetic forces will be discussed in Section 3.4.

Some examples of the applications of electric fields in DLD (eDLD) have been reported. In the first DLD paper [1], a DC electric field was used to transport DNA through a DLD device via electrophoresis and separation of kilo base pair DNA was achieved, but here the electrophoresis was seen solely as a means to drive the DNA through the array. Hanasoge et al. also used DC field to drive the flow and explore the effect of different array orientations on particle separation in eDLD [76]. The development of eDLD has been driven primarily by the Tegenfeldt group at Lund University, Sweden, [77–79] and Morgan group at University of Southampton, UK, [39,80,81], who have made use of AC fields in various modes and geometries. Among the benefits of using AC fields are the fact that the flow (carrying the particle through the device) is decoupled from the forces near to the posts (causing separation) and that much greater control can be achieved through the balancing of these forces. What is more, the nature of the force and the property of the particle that is targeted can be controlled via the frequency of the applied electric field.

In the first eDLD devices reported by Beech et al. [77], macro electrodes located inside the inlet and outlet reservoirs were used to apply an AC electric field parallel to the direction of the pressure-driven flow. The effective critical sizes were tuned in a controlled manner by up to a factor of four, for the applied voltages used. Subsequently, higher electric fields have been achieved by having electrodes right on top of an open pillar array [78] and more recently by coating pillars with platinum (3D electrode) [79] and applying the voltage between the rows of pillars. The novel 3D electrode approach obtains the closest possible inter-electrode distances, increasing the field strength per applied voltage by about three orders of magnitude. This active pillar eDLD device is able to displace particles with diameters of 250 nm, much smaller than the gap size of 11 μm. In a different approach, the Morgan group increased the field strength in eDLD by integrating microelectrodes on the sides of the pillar array. Using this configuration, Calero et al. demonstrated sorting of 500 nm from 1 and 3 μm polystyrene beads in a DLD array with a critical size of 6.3 μm using an AC voltage (50 Hz–10 kHz) [80]. In their subsequent work, by using an AC voltage with a DC bias, they were able to separate 100 nm from 500 nm and 1 μm polystyrene beads as well as sorting based on zeta potential for 3 μm microspheres [39]. Finally, in a very recent paper, Calero et al. compared their experimental data at 50 kHz with DEP numerical simulation and found good agreement [81].

Here we expand on our previous work with eDLD, using the simplest approach, i.e., macro electrodes located inside inlet and outlet reservoirs, and show tunable separation based on zeta potential of a variety of particles. We characterize and optimize our devices with regard to different running parameters using polystyrene beads with diameters from several micrometers down to 100 nanometer scales and with a wide range of zeta potentials. Finally, we apply the technique to the sorting of bio-relevant particles: Nanoscale liposomes with different sizes and zeta potentials. Liposomes have found tremendous applications in medicine and pharmacology [82] since their inception in the mid-1960s, most notably for drug delivery. What is more, liposomes are structurally and chemically similar to extracellular vesicles (EVs)—cell-derived particles involved in intercellular communication, and as such are highly relevant as a model system.

Nanoscale bio-particles, such as liposomes or EVs [83–86], have considerable potential in diagnostic and therapeutic applications. However, sorting of nanoscale particles has been identified as an important challenge [87]. Sorting of EVs down to 20 nm has been demonstrated [29] using nanofluidic DLD with feature sizes on the scale of 100 nm, but the fluidics is difficult to control and the risk of clogging is high. In addition, fabrication of these devices requires sophisticated and expensive processes like electron beam lithography, limiting broad usage. Conversely, devices with features on the micrometer scale can be fabricated using commonly available methods. We show that eDLD devices with micrometer scale gaps can be an alternative approach to the separation of nanoparticles, without the scaling down of device features and thus without some of the problems in fabrication and operation that such scaling down entails.

To optimize eDLD for different applications, it is necessary to understand how different parameters individually or mutually affect the separation. For example, increasing driving pressure can improve throughput (particle numbers or volumes), which is of great interest in many applications, but it also means a higher voltage is required to effectively displace particles. The maximum voltage that can be applied is limited by electrolysis and possibly by Joule heating through the device geometry and the conductivity of running medium. However, some particle systems place constraints on the conductivities of the suspending medium that can be used. To untwine complex dependencies as such, we scan and investigate the influence of medium conductivity, voltage, frequency, and pressure on the efficiency of separation and establish a few simple rules that may help employing eDLD for biological applications.

2. Materials and Methods

2.1. Devices and Experimental Setup

The DLD devices used throughout this paper were fabricated from polydimethylsiloxane (PDMS) using replica molding [88]. Detailed device fabrication steps can be found in Table S1 of the Supplementary Information. In each device (Figure 2a), only a narrow, single stream of sample is allowed to enter the DLD array and is buffered on two sides. Particles having different properties (size, charge) in the sample are separated as they travel along the array before entering a single outlet reservoir. At the observation windows at the beginning and at the end of the array, the lateral positions of the particles are analyzed.

Table 1. Parameters of DLD devices used in this work. Critical diameters, D_C, are nominal values based on the geometry of the pillar array (Equation (1)).

Device Name	Gap (μm)	N	D_C (μm)
Device #1	2	20	0.66
Device #2	5	10	2.32
Device #3	6	10	2.78
Device #4	8	10	3.71
Device #5	13	10	6.03
Device #6	13	5	8.41
Device #7	4	23	1.24

A list of the devices used in this work is presented in Table 1, together with their featured parameters. More details about the designs of the devices can be found in Figure S1 and Table S2 of the Supplementary Information.

During an experiment, a device is positioned on the stage of an inverted microscope (Nikon Eclipse TE2000-U, Nikon Corporation, Tokyo, Japan) (Figure 2b,c). The pressure at the inlet reservoirs (1–100 mBar) is regulated by a pressure controller (MFCS-4C, Fluigent, Paris, France). The voltage applied along the device is generated by a function generator (33120A, Hewlett Packard, Palo Alto, CA, USA) and amplified using either a high-voltage amplifier (Bipolar Operational Power Supply/Amplifier BOP 1000M, Kepco, Flushing, NY, USA) or a high-frequency amplifier (WMA-300, Falco Systems, Amsterdam, The Netherlands). The voltage is measured using an oscilloscope (54603B 60MHz, Hewlett

Packard, Palo Alto, CA, USA) with a 1×/10× probe (Kenwood PC-54, 600 V_{PP}, Havant, UK). Note that care should be taken when working with high voltages. The voltage should be turned off before any changes are made to the setup. A pair of rubber gloves should be worn at all time during the course of the experiments.

To capture experimental image stacks, a monochrome Andor Neo sCMOS camera (Andor Technology, Belfast, Northern Ireland) is used. The image stacks are analyzed using FIJI (ImageJ 1.52f, National Institutes of Health, Bethesda, MD, USA) with MOSAIC particle tracking plugin [89]. The conductivities of the media are measured with a conductivity meter (B-771 LAQUAtwin, Horiba Instruments, Kyoto, Japan). The zeta potentials of the particles are measured with a Zetasizer NanoZS instrument (Malvern Instruments, Ltd., Worcestershire, UK).

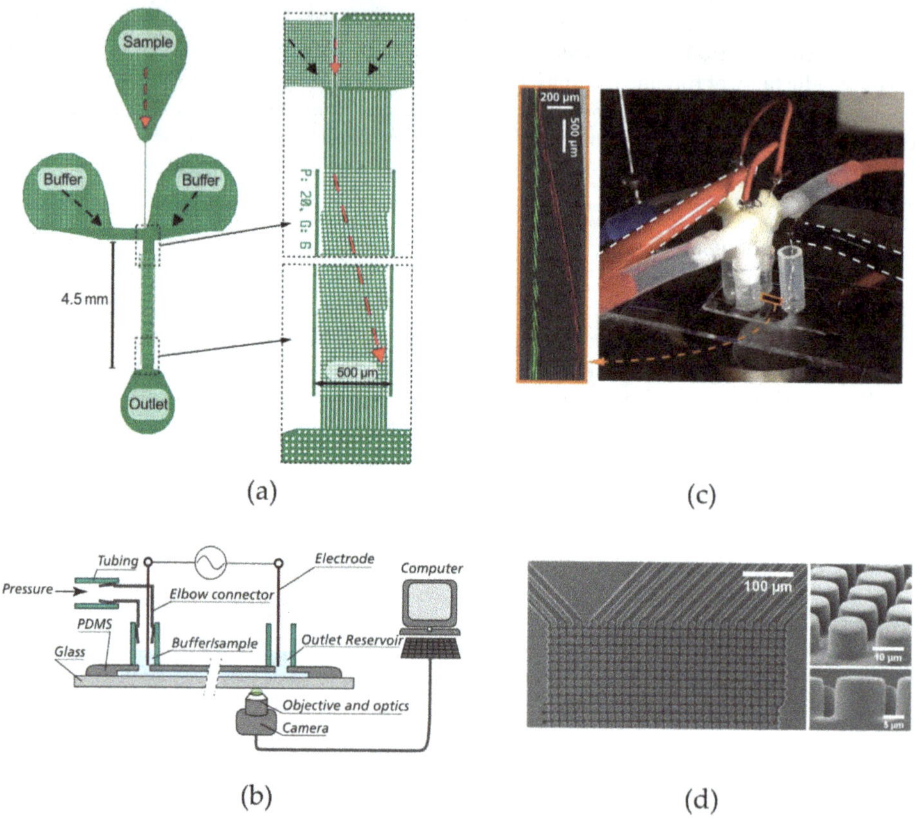

Figure 2. Devices and experimental setup. (**a**) Overview of device design. The red arrows schematically represent the trajectory of displacing particles. The actual trajectory angles of different devices are given in Table S2 of the Supplementary Information. Drawings are to scale. (**b**) Schematic of the experimental setup. (**c**) Image of the experimental setup, showing a DLD device on a microscope stage, with pressure and electrical connections. The electrical hooks are highlighted by the white dotted lines for visibility. The DLD array is enlarged and shown on the left, with color-coded trajectories of zigzagging (green) and displacing (red) particles. It has been compressed along its length by a factor of two to fit within the picture frame. (**d**) Scanning electron microscopic images of a typical DLD array (device #7, Table 1).

2.2. Data Analysis

The sorting performance of a device can be assessed by comparing lateral positions of different types of particles at the beginning and at the end of the DLD array. In our experiments, the lateral positions are discretized by the gaps. To easily compare experiments across different devices, the lateral positions can be linearly converted into percentages, with 0% displacement being perfectly zigzagging and 100% displacement being perfectly displacing. The displacement of the particles can be compared between different experiments to study the effects of different parameters including voltage, frequency, pressure, and medium conductivity. More details of the data transforming processes can be found in Figure S4 of the Supplementary Information.

To quantify the particle displacement, we need to count the numbers of particles of interest exiting at each gap at the end (outlet) of the DLD arrays. This can be done manually, particle by particle, but this method is labor intensive and inapplicable for concentrated polystyrene nanospheres or nano liposomes. Instead, provided the particle concentration is not too high, which would lead to shadowing effects, the fluorescence intensity can be used to deduce the number of particles. More specifically, in the average image of all frames of an image stack, the integrated fluorescent intensity across each gap is proportional to the number of particles passing by that gap. We denote this quantity as *inferred particle count* in the plots throughout this paper. More details of the image processing can be found in Figure S2 of the Supplementary Information and a comparison between the method and manual count can be found in Figure S3.

2.3. Sample Preparation

Particle trajectories in the DLD devices were studied for a wide range of polystyrene beads and liposomes (see more details in Table S3 of the Supplementary Information).

Polystyrene nano- and microspheres (D = 160 nm–6.3 µm), referred to as beads for brevity, were suspended in media containing KCl at various molar concentrations and 0.1% w/v Pluronic® F127 (Sigma-Aldrich Sweden AB, Stockholm, Sweden). Pluronic® was found to markedly reduce non-specific sticking of the beads to PDMS. The volumetric concentrations of the beads were 0.02% v/v for 2 µm, 160 nm, 250 nm, and 490 nm beads, and 0.086% v/v for 4.3 µm and 6.3 µm beads.

Liposomes (D ~ 150–300 nm) were made by extrusion. Two lipids with different surface charge densities, the low-charge DOPC (1,2-dioleoyl-sn-glycero-3-phosphocholine) and the high-charge POPS (1-palmitoyl-2-oleoyl-sn-glycero-3-phospho-L-serine) were mixed to achieve liposomes with different zeta potentials. DOPC lipid was mixed with a lipid conjugated Rhodamine dye (18:1 Liss Rhod PE, 1,2-dioleoyl-sn-glycero-3-phosphoethanolamine-N-(lissamine rhodamine B sulfonyl) (ammonium salt)) in chloroform at concentrations of 10 mg/mL (DOPC) and 10 µg/mL (Rhodamine) in a glass vial. POPS lipid was used at a concentration of 1 mg/mL dissolved in chloroform. The lipids and the dye were purchased from Avanti Polar Lipids (Industrial Park Drive, Alabaster, AL, USA). The lipid solutions were gently blown with nitrogen until a dry, thin film formed at the bottom of the vial. A solution of KCl (2 mM) was added to the vial to obtain concentrations of DOPC, Rhodamine, and POPS of 1 mg/mL, 1 µg/mL, and 0.1 mg/mL, respectively, and the vial was vortexed vigorously to completely dissolve the lipid film into the solution. The solution was subsequently extruded 15 times back and forth through either 100 nm or 200 nm filters on an extruder platform (T&T Scientific Corporation, 7140 Regal Lane, Knoxville, TN 37918, USA). Finally, the extruded liposomes were characterized using the Zetasizer NanoZS instrument, for both size and zeta potential, before sorting experiments. All liposome samples are monodispersed, with polydispersity indices (PDI = (std/mean)^2) below 0.24 (PDI$_{DOPC-100nm}$ = 0.10, PDI$_{DOPC-200nm}$ = 0.20, PDI$_{DOPC/POPS-100nm}$ = 0.08, PDI$_{DOPC/POPS-200nm}$ = 0.24), and no large clusters were observed in the size distributions from the DLS measurements.

3. Results and Discussion

To better understand the behavior of charged particles in an eDLD, we begin with the separation of polystyrene microspheres by their size and by their zeta potential. We then explore the capability of eDLD to effectively decrease the critical diameter to sort polystyrene nanospheres and nano-sized liposomes in micrometer-gap devices. In each case, the optimum conditions for separation regarding device geometry, applied pressure, electric field strength, and frequency are explored.

3.1. Important Sorting Parameters

We perform experiments on beads with surface carboxyl groups (4.3 μm and 6.3 μm), using device #5 with $D_C = 6.0$ μm (for 4.3 μm beads) and device #6 with $D_C = 8.4$ μm (for 6.3 μm beads) in order to determine contributions of some of the relevant sorting parameters.

Effect of frequency and medium conductivity. The applied voltage was kept constant (300 V$_{PP}$), the applied pressure varied slightly (1.0, 1.5, or 2.5 mBar) over a distance of ≈ 7 mm between inlet and outlet reservoirs and the frequency scanned in the range 1 Hz-500 kHz. Results show (Figure 3) that the 4.3 μm and 6.3 μm beads are maximally displaced at frequencies below 1 kHz for all pressures and conductivities tested. Displacement was also observed to be greater at low medium conductivity (0.7 mS/m) and lower at high conductivity (500 mS/m). Since the zeta potential is known to decrease with increasing ionic strength [90] (see Equation (6) in Table S5 for details), this indicates that the enhanced displacement due to electrokinetics increases with $|\zeta|$. For example, the absolute value of the particle zeta potential ($|\zeta|$) of 4.3 μm beads at medium conductivity of 0.7 mS/m (milliQ water), 25 mS/m (1.7 mM KCl), and 500 mS/m (37 mM KCl) are 50 mV, 24 mV, and 1 mV, respectively. For 6.3 μm beads, they are 53 mV, 27 mV, and 1 mV, respectively (Table S3 of the Supplementary Information). This will be investigated further in Section 3.2.

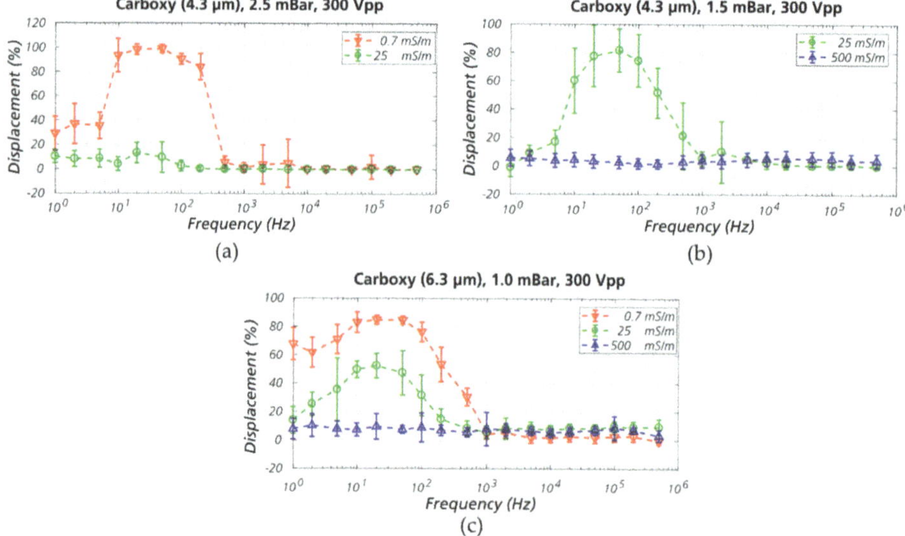

Figure 3. Effects of medium conductivity and voltage frequency on displacement of beads in eDLD. (**a,b**) The 4.3 μm beads in device #5 ($D_C = 6.0$ μm) at 2.5 mBar and 1.5 mBar, respectively. The displacement is strongest at frequencies between 10 and 500 Hz, and at low electrical conductivity of the medium. (**c**) The 6.3 μm beads in device #6 ($D_C = 8.4$ μm). The displacement is strongest at frequencies between 10 and 500 Hz, and at low conductivity.

Effect of voltage and pressure. Intuitively, the displacement increases when the applied voltage increases (the electrokinetic forces are stronger) or the applied pressure decreases (the particles travel slower, and the electrokinetic forces have more time to act). To confirm this, we kept the frequency constant at 50 Hz (which from Figure 3 gives the strongest displacement), and varied pressures from 0.5 mBar to 3 mBar in 0.5 mBar steps and applied voltages from 0 V$_{PP}$ in 10–20 V$_{PP}$ steps until the beads were completely displaced. The pressure/voltage combinations that yield 50% displacement, specifically, the threshold voltages required to displace beads to 50% at given pressures, are plotted in Figure 4.

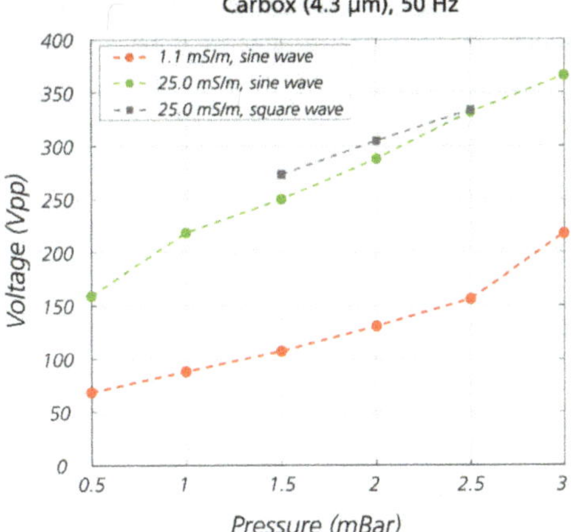

Figure 4. Effects of pressure and voltage on displacement of beads in eDLD. The data points are the voltage-pressure pairs that yield 50% displacement. More details about the data analysis can be found in Figure S4 of the Supplementary Information. The experiments were performed on 4.3 µm carboxylate beads in device #5 (D_C = 6.0 µm) at 50 Hz, in media at conductivity of 1.1 mS/m or 25 mS/m. Note that the voltages of the square wave data points have been scaled up by a factor of $\sqrt{2}$, to compare with the sine wave data points.

It is apparent from the plot that at a higher applied pressure, a higher voltage is required in order to displace particles. What is more, the plot in Figure 4 confirms our previous observation that at lower medium conductivity, the effect is stronger and thus, lower voltage is required in order to displace particles at a given pressure. Since a square waveform gives a root mean square (RMS) voltage that is $\sqrt{2}$ greater than the RMS voltage of a sine wave, we performed experiments with a square waveform. Indeed, after applying a correction term of $\sqrt{2}$ the curves corresponding to the square and the sinewave overlap. It is important to note that a square wave consists of a sum of odd harmonics making the understanding of the frequency dependence less straightforward.

3.2. Sorting Polystyrene Microspheres by Zeta Potential

To study the effect of zeta potential, polystyrene microspheres of the same size (D = 2 µm) but different surface modifications were run in device #3 (D_C = 2.8 µm). Although suspended in the same medium (25 mS/m KCl), the microspheres possess different surface charge density dependent on their types resulting in different zeta potentials (more details in Table S3 and Equation (6) of Table S5).

The experiments were performed at constant pressure and voltage (P = 10 mBar and V = 300 V_{PP}) applied over a distance of $\approx$ 7 mm between inlet and outlet reservoirs, while the frequency was varied from 10 Hz to 5 kHz. The displacement of different bead types at different frequencies are shown in Figure 5a. The distribution of displacements in Figure 5a can be fitted to normal distributions and the means are plotted against the zeta potentials (Figure 5b) and against the frequencies (Figure 5c). For illustration purposes, a supporting video showing the separation of the sulphate #2 (red) beads and carboxylate #2 (green) beads at 100 Hz can be found in the Supplementary Information (Video S1).

There is a clear correlation between the absolute value of the zeta potential and the bead displacement in Figure 5a,b. In the frequency range of 50–200 Hz, the sulphate beads, which have higher absolute values of zeta potential than that of the carboxylate beads and the amine beads, displace much more. This result demonstrates that same-sized particles with a difference in zeta potential on the order of 10 mV can be separated. Figure 5c confirms our observation in Section 3.1 that the effect is strong at low frequency ($\leq$1 kHz).

The throughput of the devices was estimated to be 2 μL/hr with a Péclet number much larger than unity. Detailed calculations are presented in Table S4 of the Supplementary Information.

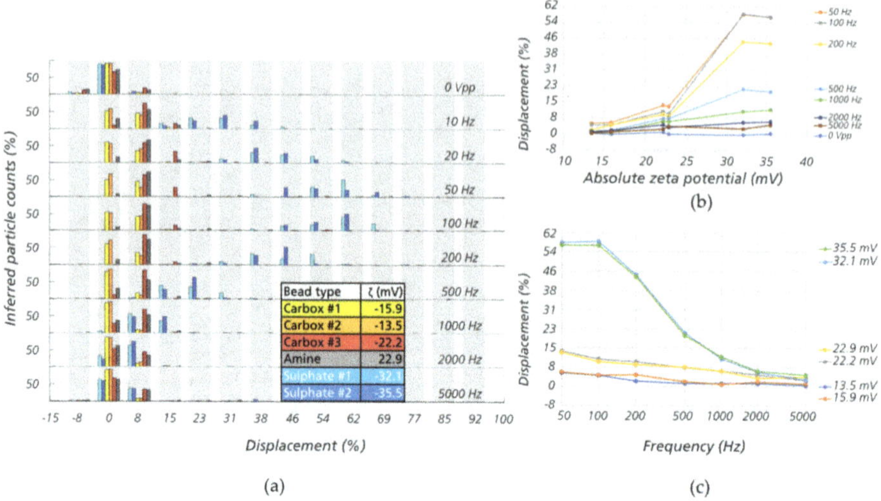

Figure 5. Displacement at outlet of 2 μm beads of different zeta potentials in device #3 (D_C = 2.8 μm). (a) Displacement of six different bead types, at frequencies from 10 to 5000 Hz. (b) Displacement as a function of zeta potential. (c) Displacement as a function of frequency. In (b,c) the standard deviation of each data point corresponds approximately to 7% displacement (one gap). Note that there is small negative displacement in plot 5a at 0 V, which can be attributed to particle diffusion (see Section 5 and Table S4 of the Supplementary Information for the calculation of diffusion length).

3.3. Sorting Polystyrene Nanospheres and Nano Liposomes

We demonstrate sorting of nano particles in devices with micrometer sized gaps by displacing 490 nm beads in device #7 (G = 4 μm, D_C = 1.24 μm) with 500 V_{PP} and 10 mBar applied over a distance of approximately ≈ 17 mm (Figure 6a). To decrease the size range further we used device #1 (G = 2 μm, D_C = 0.66 μm) to separate beads of size 160 nm and 250 nm using 1000 V_{PP} and 7.5 mBar applied over a distance of ≈ 7 mm (Figure 6b). A supporting video can be found in the Supplementary Information (Video S2).

For nano liposome separation, we used device #1 (G = 2 μm, D_C = 0.66 μm) to displace liposomes in the range 100–300 nm. We performed independent experiments on each of the four types of liposomes but in the same device and at the same voltage and pressure (500 V_{PP} and 10 mBar). The data are combined in the plots of Figure 7 for comparison. Our method allows us to separate by size (Figure 7b) or charge (Figure 7a,c) as the specific application requires. The full set of data relating to the sorting optimization of nanospheres and liposomes can be found in the Supplementary Information (Figure S5 and Figure S6, respectively). The sizes and zeta potentials of the nanospheres and nano liposomes can be found in Table S3 of the Supplementary Information.

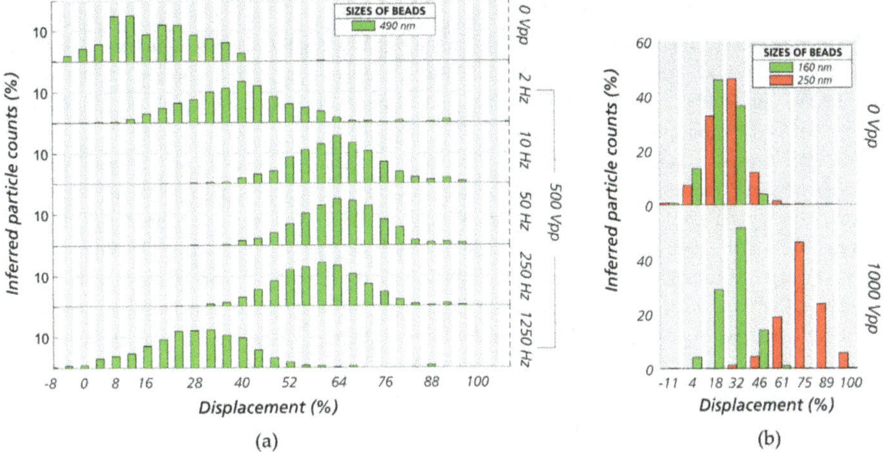

Figure 6. Displacement at outlet of nanospheres in micro-sized gap DLD devices. (**a**) Displacement of 490 nm beads in device #7 (D_C = 1.24 µm) at 500 V_{PP}. Medium: milliQ water + 0.1% Pluronic® 127 (σ = 0.5 mS/m). (**b**) Sorting of 160 nm versus 250 nm beads in device #1 (D_C = 0.66 µm) at 1000 V_{PP} at 1 kHz. Medium: milliQ water + 0.1% Pluronic® 127 (σ = 0.5 mS/m).

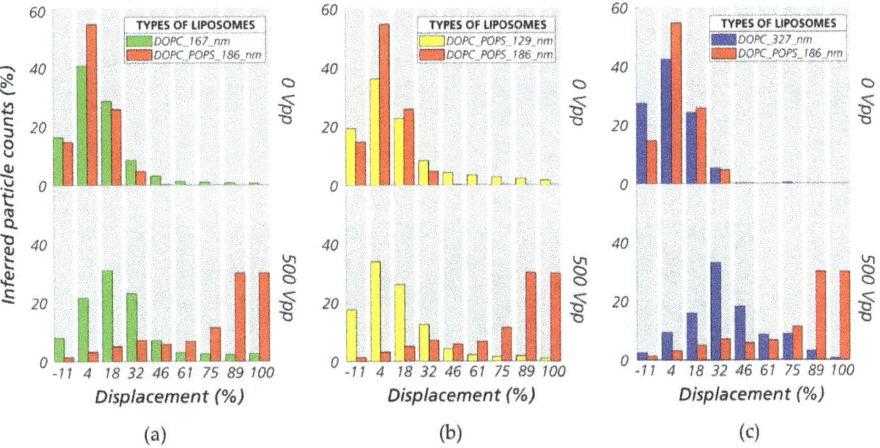

Figure 7. Sorting nano liposomes in eDLD. In general, the 1,2-dioleoyl-sn-glycero-3-phosphocholine (DOPC) liposomes possess lower charge than the DOPC+10%POPS liposomes. (**a**) Separation of liposomes of similar sizes but different surface properties ($\zeta_{DOPC,167nm}$ = −10 mV, $\zeta_{DOPC+10\%POPS,186nm}$ = −67 mV). The higher charged DOPC/1-palmitoyl-2-oleoyl-sn-glycero-3-phospho-L-serine (POPS) liposomes are more displaced. (**b**) Liposomes of different sizes but similar surface properties ($\zeta_{DOPC+10\%POPS,129nm}$ = −50 mV, $\zeta_{DOPC+10\%POPS,186nm}$ = −67 mV). The larger liposomes are more displaced. (**c**) Large liposomes with lower zeta potential ($\zeta_{DOPC,327nm}$ = −16 mV) versus small liposomes with higher zeta potential ($\zeta_{DOPC+10\%POPS,186nm}$ = −67 mV). In this case, the zeta potential is shown to be more important for the outcome of the sorting than the particle size. More information about size and zeta potential of liposomes can be found in Table S3 of the Supplementary Information. Device #1 (D_C = 0.66 µm). Frequency: 1 kHz. Medium: KCl 2 mM (σ = 29 mS/m).

3.4. The Role of Electrokinetic Driving Forces

By adding electrokinetics to DLD, its scope can be significantly broadened, which we demonstrate in the presented work. Additionally, we find the effect is only significant at low frequency ($\leq$1 kHz) and increases with zeta potential of the particles. This suggests an association with the phenomena of electrophoresis (EP) and electroosmotic flow (EOF).

Electrophoresis and electroosmotic flow refer to the motion of charged particles and electrolyte solution, respectively, under an electric field. A charged particle suspended in an electrolyte solution will travel under the application of an electric field (particle EP). At the same time, the electrolyte solution contained in a channel with charged walls also moves when the field is on (EOF). In fact, particle EP stems from the EOF near the surface of the particle, so the two phenomena are related [41,43,91]. On a global scale, the total electrokinetic mobility of the particle is the algebraic sum of the electrophoretic mobility and the electroosmotic mobility. Under low-frequency AC fields, the particles can be seen oscillating along the direction of the field, as shown in Video S8 of the Supplementary Information.

When a pressure is applied together with a low-frequency AC field, the particle can be seen traveling in the direction of the pressure gradient but at the same time oscillating at the frequency of the applied field. This can be seen in Video S4 of the Supplementary Information (pressure driven flow + AC 50 Hz), as compared to Video S3, where only a pressure was applied. The electrokinetic wall force depicted in Figure 1c can be observed in Video S5 (pressure driven flow + AC 10 Hz field) at time 0.34 s when the particle is "pushed" away from the pillar, leading to enhanced displacement. Note that the direction of this movement is perpendicular to the pillar and is different from the direction of the EP/EOF oscillation, which is tangential to the pillar (because the pillar is insulating, the electric field is tangential to the pillar). In Figure 8, trajectories of 2 µm carboxylate beads #1 at different frequencies are compared. It can be seen clearly that starting from 200 Hz, the beads displace less, which is consistent with the low-frequency operating range we have reported in Sections 3.1 and 3.2.

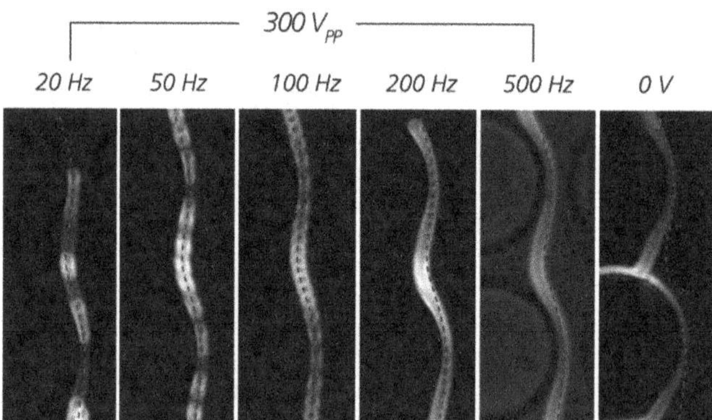

Figure 8. Comparison of trajectories of 2 µm carboxylate beads #1 at different frequencies. Identical red dotted lines based on trajectory at 100 Hz have been added to the images to guide the eye. Device #2 (D_C = 2.3 µm) was used. The applied pressure was 30 mBar. The medium was KCl 22 mS/m with 2.5% *w/v* polyvinylpyrrolidone (PVP). Similar to Pluronic® 127, PVP helps reduce non-specific adsorption of the beads to the PDMS walls of the device.

At very low frequency ($\leq$10 Hz), although the electrokinetic wall force might also be strong, the large oscillation magnitude of the particles may hinder the displacement. This can be observed in Video S6 (pressure driven flow + AC 2 Hz). During the negative phase of the oscillation, at around 1.00 sec, the particle is pulled backwards in the opposite direction to the pressure driven flow and may cross streamlines in the opposite direction to the displacement direction. This leads to particles moving in the zigzag mode while having a strong displacing tendency otherwise. This explains the lower displacement of particles at very low frequencies in the plots of Figure 3.

The prospects of various electrokinetic mechanisms influencing sorting in DLD devices have been brought up for discussion in the literature [80,81]. Due to viscous damping, EOF and EP is not expected to be of importance at higher frequencies. Calero et al. [80] identified 500 Hz as a reasonable threshold frequency, above which EOF and EP would not be important.

It is at this point relevant to mention that there are two types of EOF: *Linear* and *non-linear EOF* [92]. In linear EOF, the external field is much smaller than the field in the electric double layer (EDL) (which can be estimated by the zeta potential and the Debye length, and is on the order of 10^6 V/m). The external field therefore does not affect the distribution of charge in the EDL. With this condition, the velocity of the fluid or a particle can be calculated by Smoluchowski's equation:

$$v = \frac{\zeta \varepsilon}{\eta} E \tag{2}$$

where ζ is the zeta potential of the wall or particle, ε is the permittivity of the fluid, η is the viscosity of the fluid, and E is the external field. If $\zeta > 0$, v is parallel with E for EP and anti-parallel with E for EOF.

On the contrary, if the external field gives rise to or modifies the charge distribution in the EDL, ζ in Equation (2) is no longer independent of E and the velocity no longer varies linearly with the external field. There are several embodiments of this non-linear electrokinetics reported in the literature. For instance, when the external field polarizes conductive walls (metal electrodes), it exerts at the same time an electric force on the newly induced charge and causes circulation of fluid just above the electrodes. This phenomenon is called *AC-electroosmosis* [93–95] and the time-averaged electroosmotic velocity is proportional to the square of the applied voltage [96]:

$$\langle v \rangle \propto \frac{\varepsilon}{\eta} f(\omega, \varepsilon, \sigma, \kappa, z) V^2 \tag{3}$$

where f is a function of angular frequency ω, medium permittivity ε, medium conductivity σ, inverse of Debye length κ, and distance z from electrodes.

Similarly, for the case of conducting particles (or in 2D, a rod), one can observe vortices near the particle surface. This phenomenon is called *induced-charge electroosmosis/electrophoresis (ICEO/ICEP)* [97,98] and the typical flow speeds are also proportional to the square of the applied field:

$$U_0 \propto \frac{\varepsilon}{\eta} a E^2 \tag{4}$$

Non-linear electrokinetics for dielectrics, which is more relevant to our work, has also been studied [99,100]. Notably, Wang et al. [100] observed vortices of 1 μm polystyrene particles near a narrow constriction when a 1 kHz AC field was applied, which they attributed to ICEO and electrothermal flow (the former dominated in low conductivity medium). We also observe vortices of 1.1 μm polystyrene particles suspended in milliQ water in a pillar array under 100 Hz AC field (Figure 9 and Supplementary Information Video S7) when no pressure driven flow is present. Similar vortices have also been reported by Calero et al. [81] in their type of eDLD with microelectrodes on the sides of the device.

In our eDLD experiments, since the enhanced displacement effect (from now on called "the effect") decreases with increasing medium conductivity at any given applied voltage, electrothermal flow can be excluded. Since the effect is significant at low frequency and increases with the zeta potential, it may have some dependence on EP and EOF (either linear or non-linear). To investigate this, we look into the displacement of 4.3 μm beads at different applied pressures and voltages, and at two different conductivities (Figure 10). For each conductivity, the plot on the left shows the displacement as a function of the voltage. These plots confirm that at higher pressure, a higher voltage is required to achieve the same displacement as in the low-pressure case. This is because the particles travel faster, and the electrokinetic forces have less time to act at higher pressure.

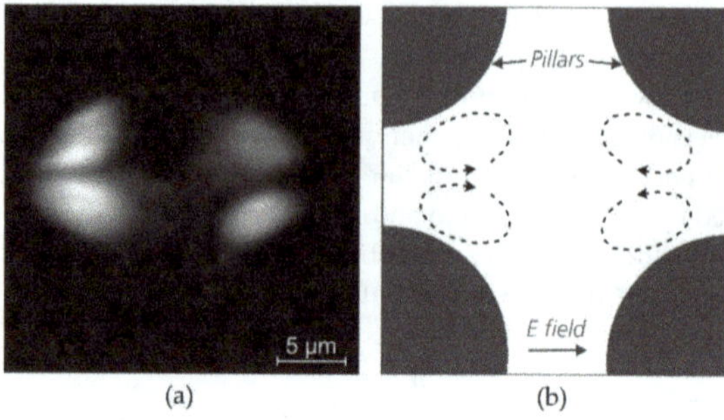

(a)   (b)

Figure 9. Vortices of 1.1 µm polystyrene beads in milliQ water at 500 V_{PP} at 100 Hz, in device #4 (G = 8 µm, distance between electrodes ≈ 7 mm). (**a**) Actual image, (**b**) Diagram showing the direction of the vortices). The original image stack (*557* frames @25fps) has the field of view of 8 × 7 identical unit cells (the image shows one unit cell). To simultaneously observe the dynamics of many particles, we used MATLAB 2017a (Mathworks, Natick, MA, USA) to superimpose all these unit cells into one unit cell. The resulted image stack was then averaged to produce the image shown in the figure. A video with particle tracking is available in the Supplementary Information (Video S7).

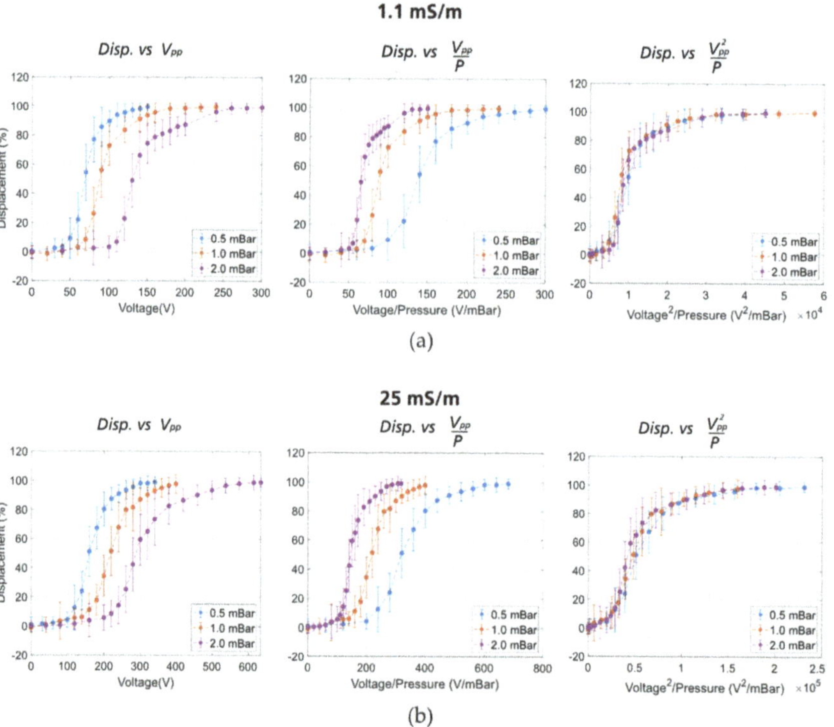

Figure 10. Displacement of 4.3 µm beads as a function of voltage, voltage/pressure, or voltage²/pressure. (**a**) At medium conductivity of 1.1 mS/m. (**b**) At medium conductivity of 25 mS/m.

To compare different experiments, the pressure can be normalized (the middle plot for each conductivity in Figure 10). In these plots, if the effect is linearly dependent on the voltage, the data at different pressures should collapse into one curve (see Section 6 in the Supplementary Information for a simplified mathematical model). It is shown not to be the case. However, when we plot the displacement as a function of voltage squared over pressure (the plot on the right for each conductivity of Figure 10), the data collapse to one curve within the uncertainty of our experiments. This suggests association with non-linear electrokinetics, which varies with the square of the applied voltage, as shown in Equations (3) and (4).

It is necessary to discuss dielectrophoresis (DEP) as a possible mechanism since the DEP force is also proportional to V^2. The DEP force on a spherical particle of radius r in a non-uniform electric field E is given by:

$$F_{DEP} = 2\pi\varepsilon_m r^3 Re\left(\frac{\widetilde{\varepsilon}_p - \widetilde{\varepsilon}_m}{\widetilde{\varepsilon}_p + 2\widetilde{\varepsilon}_m}\right)\nabla|E_{rms}|^2 \tag{5}$$

where E_{rms} is the root mean square value of the field, $\widetilde{\varepsilon}_p$, $\widetilde{\varepsilon}_m$ are the complex permittivity of the particle (p) and the suspending medium (m) and are defined as: $\widetilde{\varepsilon} = \varepsilon - i\sigma/\omega$. Here ε is the permittivity, σ is the conductivity, and ω is the angular frequency of the electric field. The term in the brackets represents the relative contrast between the permittivities of the particle and the medium and is called the Clausius–Mossotti factor: $f_{CM} = \frac{\widetilde{\varepsilon}_p - \widetilde{\varepsilon}_m}{\widetilde{\varepsilon}_p + 2\widetilde{\varepsilon}_m}$. For polystyrene beads, the bulk conductivity σ_b is negligibly small and their conductivity comes from the surface conductivity σ_s [101]:

$$\sigma_p \approx \sigma_s = \frac{2K}{r} \tag{6}$$

where K is the surface conductance of the beads. For non-conducting polystyrene particles, typically $K \sim 1nS$ [102,103]. According to Equation (5), the polarity of the DEP force is dependent on the real part of the Clausius–Mossotti factor. When $Re(f_{CM}) > 0$, the particles travel to the regions of high field strength (positive or pDEP), conversely when $Re(f_{CM}) < 0$, the particles travel to the regions of low field strength (negative or nDEP). In eDLD with the electric field parallel to the flow field, as has been pointed out by Beech et al. [77], the nDEP force "pushes" the particles away from the pillar out of the first stream and, if sufficiently strong, into the subsequent stream (i.e., in same direction as F_{EW} in Figure 1c). Conversely, pDEP "pulls" the particles towards the pillar, which intuitively reduces the effect. To assess the contribution of DEP, the real parts of the Clausius–Mossotti factors for 2 μm and 4.3 μm polystyrene beads at different conductivities used in this paper are plotted in Figure 11.

Figure 11 shows that at high frequencies ($\geq$ 1 MHz), the CM factor approaches a value close to -0.5 asymptotically, giving rise to nDEP. At lower frequencies it is positive (pDEP) for small particles in low conductivity medium (here only for 2 μm beads in milliQ water) and nDEP for large particles or high medium conductivities relative to the particle conductivity. In our experimental data presented in Sections 3.1 and 3.2, we observe a significantly enhanced displacement effect only at frequencies between 10 Hz and 1 kHz (the green region in Figure 11) whereas the DEP force, whether it is pDEP or nDEP, is constant in a much broader frequency range, from 10 Hz to 100 kHz. This indicates that DEP is not the main contribution to the effect and suggests that non-linear EOF or EP probably plays a more important role. It should be noted that nDEP can be utilized to enhance particle displacement but required applied voltages might be higher than those used here.

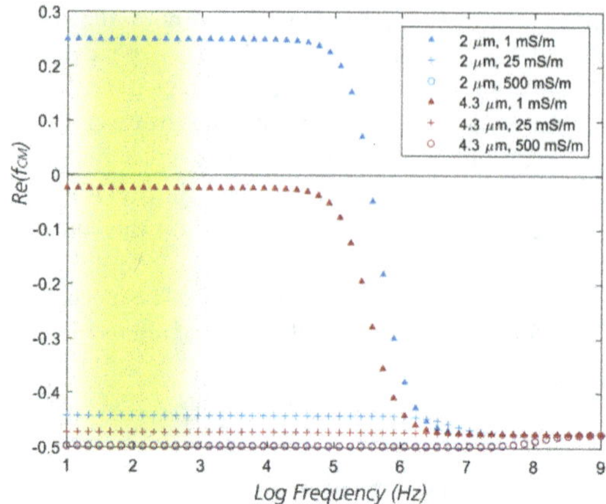

Figure 11. The real part of the Clausius–Mossotti factor for 2 µm and 4.3 µm beads at different conductivities used in the experiments, assuming $K = 1$ nS for all beads. The blurred green region highlights the range of frequencies where we observe the enhanced displacement effect.

Finally, we consider whether or not it is possible to improve future devices based on the findings of the present study. The question can only be answered in relation to the specific requirements of the application, but we can give some general guidelines. Larger devices can provide higher throughput. However, higher applied voltages are required in order to maintain the necessary field strengths inside pillar arrays, which constitutes a practical upper limit on device size. It is possible to increase the field gradients inside a pillar array for a given applied voltage, which can improve device performance, but these improvements also come at a cost. For example, simulations (see Figure S8c of the Supplementary Information) show that changing the gap (G) and the period (N) such that the critical size remains the same does not have a significant effect on the field strength at one critical radius distance from the posts, and we conclude that these parameters alone cannot be used to greatly improve the voltage response in the device. Changing the diameters (P) of the posts however has a larger impact on the local field strength for a given applied voltage (see Figure S8d of the Supplementary Information). This can be understood by considering the ratio P/G, which indicates the amount by which the electric field is constricted between the posts. Although the field strength at one critical radius distance from the posts increases 35% when P/G goes from 0.28 to 3.3 with Dc kept constant, which is significant, the device would also need to be considerably longer, 3.4 times longer in this case, which entails a large increase in the required applied voltage. Since large N also leads to longer devices for a given critical size, in general, small N and large P/G lead to higher fields for a given applied voltage. Other changes to both the geometry of the post array and to the shapes of the posts are expected to change the distribution of the electric field in the post array and could also be used to optimize device performance for specific applications, but that is outside the scope of the present work and will be explored in future studies.

4. Conclusions

We have presented an integration of DLD and electrokinetics to sort particles based not only on size, but also on zeta potential which in turn depends on the surface charge. The integration has been done in the simplest way possible, with electrodes located in inlet and outlet reservoirs. Without having to fabricate microelectrodes or scale down the devices, we were able to sort nanoparticles in devices having micrometer-sized gaps, in a way that is sensitive to size and zeta potential. Our approach is applicable to all particle sizes from the micro- down to the nanoscale and is not bound by requirements for very low salt concentrations for the running medium.

We posit that non-linear electroosmotic flow and electrophoresis play an important role in the mechanism of zeta-potential-based or charge-based sorting using eDLD. However, further studies are necessary to fully understand how these mechanisms interact and are responsible for the separation. Nevertheless, we have demonstrated that by optimizing running parameters such as ionic strength, applied voltage, frequency, and pressure, we can adapt the approach to a wide range of relevant particle sizes and zeta potentials.

Our results prove that eDLD can be used to sort nano particles such as liposomes with strong biological relevance, for example in medicine and pharmacology, most notably for drug delivery. This opens up for sorting extracellular vesicles (EVs), which are similar in structure to liposomes. EVs have great potential in diagnostics and therapeutics but current EV manufacturing and sorting methods are inadequate in terms of purity, reproducibility, yield, time, and cost [84,85]. From a fabrication perspective our eDLD devices are simple, disposable, and can be realized in cheap materials using standard techniques. Ongoing work is focused on using what we have learned to developing devices that will be capable of sorting EVs based on size and surface charge, at throughputs relevant for further bioanalysis.

Supplementary Materials: The following are available online at http://www.mdpi.com/2072-666X/11/11/1014/s1: Figure S1: Overview of a typical device, Figure S2: Image processing steps used to estimate relative particle counts based on fluorescence intensity, Figure S3: Relative amounts of 2 μm beads as measured by manual counting and by integrated fluorescent intensity, Figure S4: Summary of data transforming steps in order to investigate the effects of different running parameters on particle displacement, Figure S5: Optimization of nanosphere sorting, Figure S6: Optimization of liposome sorting, with voltage as the changing parameter, Figure S7: Simple sketch demonstrating the displacement caused by electrokinetic force, Figure S8: Electric field simulation of eDLD, Table S1: Device fabrication steps, Table S2: Parameters of DLD devices used in this work, Table S3: Specifications of the particles used in this work, Table S4: Throughput and Péclet numbers of the experiments reported in the main text, Table S5: Important equations of electrokinetics, Video S1: Sorting—Sulphate2 (red)—Carbox2 (green)—100 Hz. *Real-time, color-coded video showing the separation of the sulphate #2 (red) and carboxylate #2 (green) beads (both ≈ 2μm) in device #3 (D_C = 2.8 μm) when the electric field was turned on at the middle of the video. The result is plotted in Figure 5 (main text).* Video S2: Sorting—Red 250 nm—Green 160 nm—0 V_{PP} then 1000 V_{PP}—1000 Hz. *Sorting of 250 nm and 160 nm beads at an applied voltage of 1000 V_{PP} at 1 kHz. The separation is possible only after the voltage was turned on. The playback is fast-forwarded twice.* Video S3: Trajectory—4um3 particle under pressure driven flow. *Trajectories of of a 4.3 μm bead in device #5 (D_C = 6.0 μm), under a pressure of 1.5 mBar, in KCl 25 mS/m with 0.1% w/v Pluronic® 127. The pressure driven flow is from left to right. The playback is slowed down 10x.* Video S4–S6: Similar to Video S3 but with voltage of 300 V_{PP}/square wave at frequencies of 50 Hz, 10 Hz, and 2 Hz, respectively. *The playback is slowed down 40x, 40x, and 10x, respectively. The pressure driven flow is from left to right. The oscillatory motion of beads can be seen clearly in Video S4 (50 Hz) and Video S5 (10 Hz). The enhanced displacement can be seen clearly in Video S5 (10 Hz) at 0.34 s. The setback from displacing to zigzag at very low frequency can be seen in Video S6 (2 Hz) at 1.00 sec. Note that we used square wave in these slow-motion videos, instead of the sine wave as in most of the data in the main text, to give the eyes an easier detection of the phase of the oscillation.* Video S7: Trajectory—1um1 particles under 100 Hz AC signal (no pressure driven flow). *Vortexes of 1.1 μm beads in milliQ water at 500 V_{PP} at 100 Hz, in device #4 (G = 8 μm). The image is combined from 557 frames @25fps of 8 × 4 different unit cells (i.e., the full field of view is 32 times as wide as the video dimensions).* Video S8: Oscillation—2um particles—100Hz (red) 500Hz (green) 5000Hz (magenta)—(superimposed) .mp4 (no pressure driven flow). *Superimposed, color-coded video showing the oscillatory motion of the same type of bead at different frequencies, in the gap between two pillars, when an AC voltage was applied from left to right (no pressure applied). The video was combined from three videos at three different frequencies. The video has been slowed down by 50 times.*

Author Contributions: Conceptualization, all authors; methodology, all authors; software, B.D.H.; validation, all authors; formal analysis, B.D.H.; investigation, B.D.H.; resources, J.O.T.; data curation, B.D.H.; writing—original draft preparation, B.D.H.; writing—review and editing, all authors; visualization, B.D.H. and J.P.B.; supervision, J.O.T. and J.P.B.; project administration, J.O.T.; funding acquisition, J.O.T. All authors have read and agreed to the published version of the manuscript.

Funding: This research was funded by the European Union, grant number 607350 (project LAPASO), grant number 801367 (project evFOUNDRY), grant number 634890 (project BeyondSeq), by the Swedish Research council, grant number 2016-05739 and NanoLund. All device processing was conducted within Lund Nano Lab.

Acknowledgments: We are thankful to Hywel Morgan, Carlos Honrado, Daniel Spencer, and Nicolas Green (University of Southampton) for their generous sharing of expertise in electrokinetics. We are grateful to Tommy Nylander (Lund University) for his help with making the liposomes.

Conflicts of Interest: The authors declare no conflict of interest. The funders had no role in the design of the study, in the collection, analyses, or interpretation of data, in the writing of the manuscript, or in the decision to publish the results.

References

1. Huang, L.R.; Cox, E.C.; Austin, R.H.; Sturm, J.C. Continuous particle separation through deterministic lateral displacement. *Science* **2004**, *304*, 987–990. [CrossRef]
2. Siyang, Z.; Yung, R.; Yu-Chong, T.; Kasdan, H. Deterministic lateral displacement MEMS device for continuous blood cell separation. In Proceedings of the 18th IEEE International Conference on Micro Electro Mechanical Systems, Miami Beach, FL, USA, 30 January–3 February 2005; pp. 851–854.
3. Davis, J.A.; Inglis, D.W.; Morton, K.J.; Lawrence, D.A.; Huang, L.R.; Chou, S.Y.; Sturm, J.C.; Austin, R.H. Deterministic hydrodynamics: Taking blood apart. *Proc. Natl. Acad. Sci. USA* **2006**, *103*, 14779–14784. [CrossRef]
4. Li, N.; Kamei, D.T.; Ho, C.M. On-chip continuous blood cell subtype separation by deterministic lateral displacement. In Proceedings of the 2007 2nd IEEE International Conference on Nano/Micro Engineered and Molecular Systems, Bangkok, Thailand, 16–19 January 2007; pp. 932–936.
5. Inglis, D.W.; Lord, M.; Nordon, R.E. Scaling deterministic lateral displacement arrays for high throughput and dilution-free enrichment of leukocytes. *J. Micromechanics Microengineering* **2011**, *21*, 054024. [CrossRef]
6. Holmes, D.; Whyte, G.; Bailey, J.; Vergara-Irigaray, N.; Ekpenyong, A.; Guck, J.; Duke, T. Separation of blood cells with differing deformability using deterministic lateral displacement. *Interface Focus* **2014**, *4*, 20140011. [CrossRef]
7. Loutherback, K.; D'Silva, J.; Liu, L.; Wu, A.; Austin, R.H.; Sturm, J.C. Deterministic separation of cancer cells from blood at 10 mL/min. *AIP Adv.* **2012**, *2*, 42107. [CrossRef]
8. Liu, Z.; Huang, F.; Du, J.; Shu, W.; Feng, H.; Xu, X.; Chen, Y. Rapid isolation of cancer cells using microfluidic deterministic lateral displacement structure. *Biomicrofluidics* **2013**, *7*, 11801. [CrossRef]
9. Karabacak, N.M.; Spuhler, P.S.; Fachin, F.; Lim, E.J.; Pai, V.; Ozkumur, E.; Martel, J.M.; Kojic, N.; Smith, K.; Chen, P.I.; et al. Microfluidic, marker-free isolation of circulating tumor cells from blood samples. *Nat. Protoc.* **2014**, *9*, 694–710. [CrossRef] [PubMed]
10. Okano, H.; Konishi, T.; Suzuki, T.; Suzuki, T.; Ariyasu, S.; Aoki, S.; Abe, R.; Hayase, M. Enrichment of circulating tumor cells in tumor-bearing mouse blood by a deterministic lateral displacement microfluidic device. *Biomed. Microdevices* **2015**, *17*, 9964. [CrossRef] [PubMed]
11. Au, S.H.; Edd, J.; Stoddard, A.E.; Wong, K.H.K.; Fachin, F.; Maheswaran, S.; Haber, D.A.; Stott, S.L.; Kapur, R.; Toner, M. Microfluidic isolation of circulating tumor cell clusters by size and asymmetry. *Sci. Rep.* **2017**, *7*, 2433. [CrossRef] [PubMed]
12. Liu, Z.; Chen, R.; Li, Y.; Liu, J.; Wang, P.; Xia, X.; Qin, L. Integrated microfluidic chip for efficient isolation and deformability analysis of circulating tumor cells. *Adv. Biosyst.* **2018**, *2*, 1800200. [CrossRef]
13. Holm, S.H.; Beech, J.P.; Barrett, M.P.; Tegenfeldt, J.O. Separation of parasites from human blood using deterministic lateral displacement. *Lab Chip* **2011**, *11*, 1326–1332. [CrossRef] [PubMed]
14. Holm, S.H.; Beech, J.P.; Barrett, M.P.; Tegenfeldt, J.O. Simplifying microfluidic separation devices towards field-detection of blood parasites. *Anal. Methods* **2016**, *8*, 3291–3300. [CrossRef]
15. Huang, R.; Barber, T.A.; Schmidt, M.A.; Tompkins, R.G.; Toner, M.; Bianchi, D.W.; Kapur, R.; Flejter, W.L. A microfluidics approach for the isolation of nucleated red blood cells (NRBCs) from the peripheral blood of pregnant women. *Prenat. Diagn.* **2008**, *28*, 892–899. [CrossRef]
16. Laki, A.J.; Botzheim, L.; Ivan, K.; Szabo, T.; Tamasi, V.; Buzas, E.; Civera, P. Microvesicle fractionation using deterministic lateral displacement effect. In Proceedings of the 9th IEEE International Conference on Nano/Micro Engineered and Molecular Systems (NEMS), Waikiki Beach, HI, USA, 13–16 April 2014; pp. 490–493.
17. Laki, A.J.; Botzheim, L.; Ivan, K.; Tamasi, V.; Civera, P. Separation of microvesicles from serological samples using deterministic lateral displacement effect. *Bionanoscience* **2015**, *5*, 48–54. [CrossRef]
18. D'Silva, J.; Austin, R.H.; Sturm, J.C. Inhibition of clot formation in deterministic lateral displacement arrays for processing large volumes of blood for rare cell capture. *Lab Chip* **2015**, *15*, 2240–2247. [CrossRef] [PubMed]

19. Green, J.V.; Radisic, M.; Murthy, S.K. Deterministic lateral displacement as a means to enrich large cells for tissue engineering. *Anal. Chem.* **2009**, *81*, 9178–9182. [CrossRef] [PubMed]

20. Zhang, B.; Green, J.V.; Murthy, S.K.; Radisic, M. Label-free enrichment of functional cardiomyocytes using microfluidic deterministic lateral flow displacement. *PLoS ONE* **2012**, *7*, e37619. [CrossRef] [PubMed]

21. Liu, Z.; Lee, Y.; Jang, J.; Li, Y.; Han, X.; Yokoi, K.; Ferrari, M.; Zhou, L.; Qin, L. Microfluidic cytometric analysis of cancer cell transportability and invasiveness. *Sci. Rep.* **2015**, *5*, 14272. [CrossRef] [PubMed]

22. Tottori, N.; Nisisako, T.; Park, J.; Yanagida, Y.; Hatsuzawa, T. Separation of viable and nonviable mammalian cells using a deterministic lateral displacement microfluidic device. *Biomicrofluidics* **2016**, *10*. [CrossRef]

23. Xavier, M.; Holm, S.H.; Beech, J.P.; Spencer, D.; Tegenfeldt, J.O.; Oreffo, R.O.C.; Morgan, H. Label-free enrichment of primary human skeletal progenitor cells using deterministic lateral displacement. *Lab Chip* **2019**, *19*, 513–523. [CrossRef]

24. Inglis, D.W.; Herman, N.; Vesey, G. Highly accurate deterministic lateral displacement device and its application to purification of fungal spores. *Biomicrofluidics* **2010**, *4*. [CrossRef] [PubMed]

25. Joensson, H.N.; Uhlen, M.; Svahn, H.A. Droplet size based separation by deterministic lateral displacement-separating droplets by cell—Induced shrinking. *Lab Chip* **2011**, *11*, 1305–1310. [CrossRef] [PubMed]

26. Tottori, N.; Hatsuzawa, T.; Nisisako, T. Separation of main and satellite droplets in a deterministic lateral displacement microfluidic device. *RSC Adv.* **2017**, *7*, 35516–35524. [CrossRef]

27. Beech, J.P.; Ho, B.D.; Garriss, G.; Oliveira, V.; Henriques-Normark, B.; Tegenfeldt, J.O. Separation of pathogenic bacteria by chain length. *Anal. Chim. Acta* **2018**, *1000*, 223–231. [CrossRef]

28. Wunsch, B.H.; Kim, S.C.; Gifford, S.M.; Astier, Y.; Wang, C.; Bruce, R.L.; Patel, J.V.; Duch, E.A.; Dawes, S.; Stolovitzky, G.; et al. Gel-on-a-chip: Continuous, velocity-dependent DNA separation using nanoscale lateral displacement. *Lab A Chip* **2019**, *19*, 1567–1578. [CrossRef]

29. Wunsch, B.H.; Smith, J.T.; Gifford, S.M.; Wang, C.; Brink, M.; Bruce, R.L.; Austin, R.H.; Stolovitzky, G.; Astier, Y. Nanoscale lateral displacement arrays for the separation of exosomes and colloids down to 20 nm. *Nat. Nanotechnol.* **2016**, *11*, 936–940. [CrossRef]

30. McGrath, J.; Jimenez, M.; Bridle, H. Deterministic lateral displacement for particle separation: A review. *Lab Chip* **2014**, *14*, 4139–4158. [CrossRef]

31. Salafi, T.; Zhang, Y.; Zhang, Y. A review on deterministic lateral displacement for particle separation and detection. *Nano Micro Lett.* 2019; 11, 77. [CrossRef]

32. Zeming, K.K.; Thakor, N.V.; Zhang, Y.; Chen, C.H. Real-time modulated nanoparticle separation with an ultra-large dynamic range. *Lab Chip* **2016**, *16*, 75–85. [CrossRef]

33. Mutlu, B.R.; Smith, K.C.; Edd, J.F.; Nadar, P.; Dlamini, M.; Kapur, R.; Toner, M. Non-equilibrium inertial separation array for high-throughput, large-volume blood fractionation. *Sci. Rep.* **2017**, *7*, 9915. [CrossRef]

34. Frechette, J.; Drazer, G. Directional locking and deterministic separation in periodic arrays. *J. Fluid Mech.* **2009**, *627*, 379–401. [CrossRef]

35. Davis, J.A. Microfluidic Separation of Blood Components through Deterministic Lateral Displacement. Ph.D. Thesis, Princeton University, Princeton, NJ, USA, 2008.

36. Beech, J.P.; Holm, S.H.; Adolfsson, K.; Tegenfeldt, J.O. Sorting cells by size, shape and deformability. *Lab Chip* **2012**, *12*, 1048–1051. [CrossRef] [PubMed]

37. Jiang, M.L.; Budzan, K.; Drazer, G. Fractionation by shape in deterministic lateral displacement microfluidic devices. *Microfluid. Nanofluidics* **2015**, *19*, 427–434. [CrossRef]

38. Henry, E.; Holm, S.H.; Zhang, Z.; Beech, J.P.; Tegenfeldt, J.O.; Fedosov, D.A.; Gompper, G. Sorting cells by their dynamical properties. *Sci. Rep.* **2016**, *6*, 34375. [CrossRef] [PubMed]

39. Calero, V.; Garcia-Sanchez, P.; Ramos, A.; Morgan, H. Combining DC and AC electric fields with deterministic lateral displacement for micro- and nano-particle separation. *Biomicrofluidics* **2019**, *13*, 054110. [CrossRef]

40. Henry, D. The cataphoresis of suspended particles. Part I.—The equation of cataphoresis. *Proc. R. Soc. Lond. Ser. A Contain. Pap. A Math. Phys. Character* **1931**, *133*, 106–129.

41. Wiersema, P.H.; Loeb, A.L.; Overbeek, J.T.G. Calculation of the electrophoretic mobility of a spherical colloid particle. *J. Colloid Interface Sci.* **1966**, *22*, 78–99. [CrossRef]

42. Morrison Jr, F.A. Electrophoresis of a particle of arbitrary shape. *J. Colloid Interface Sci.* **1970**, *34*, 210–214. [CrossRef]

43. Viovy, J.L. Electrophoresis of DNA and other polyelectrolytes: Physical mechanisms. *Rev. Mod. Phys.* **2000**, *72*, 813–872. [CrossRef]

44. Pohl, H.A. *Dielectrophoresis: The Behavior of Neutral Matter in Nonuniform Electric Fields (Cambridge Monographs on Physics)*; Cambridge University Press: Cambridge, UK; New York, NY, USA, 1978.

45. Pethig, R. Review article-dielectrophoresis: Status of the theory, technology, and applications. *Biomicrofluidics* **2010**, *4*. [CrossRef]

46. Pethig, R.R. *Dielectrophoresis: Theory, Methodology and Biological Applications*; Wiley: New Jersey, NJ, USA, 2017.

47. Pohl, H.A.; Hawk, I. Separation of living and dead cells by dielectrophoresis. *Science* **1966**, *152*, 647–649. [CrossRef]

48. Markx, G.H.; Talary, M.S.; Pethig, R. Separation of viable and non-viable yeast using dielectrophoresis. *J. Biotechnol.* **1994**, *32*, 29–37. [CrossRef]

49. Markx, G.H.; Pethig, R. Dielectrophoretic separation of cells: Continuous separation. *Biotechnol. Bioeng* **1995**, *45*, 337–343. [CrossRef] [PubMed]

50. Church, C.; Zhu, J.; Wang, G.; Tzeng, T.R.; Xuan, X. Electrokinetic focusing and filtration of cells in a serpentine microchannel. *Biomicrofluidics* **2009**, *3*, 44109. [CrossRef] [PubMed]

51. Kang, Y.; Li, D.; Kalams, S.A.; Eid, J.E. DC-Dielectrophoretic separation of biological cells by size. *Biomed. Microdevices* **2008**, *10*, 243–249. [CrossRef] [PubMed]

52. Gallo-Villanueva, R.C.; Jesus-Perez, N.M.; Martinez-Lopez, J.I.; Pacheco, A.; Lapizco-Encinas, B.H. Assessment of microalgae viability employing insulator-based dielectrophoresis. *Microfluid. Nanofluidics* **2011**, *10*, 1305–1315. [CrossRef]

53. Lapizco-Encinas, B.H.; Simmons, B.A.; Cummings, E.B.; Fintschenko, Y. Dielectrophoretic concentration and separation of live and dead bacteria in an array of insulators. *Anal. Chem.* **2004**, *76*, 1571–1579. [CrossRef]

54. Lapizco-Encinas, B.H.; Simmons, B.A.; Cummings, E.B.; Fintschenko, Y. Insulator-based dielectrophoresis for the selective concentration and separation of live bacteria in water. *Electrophoresis* **2004**, *25*, 1695–1704. [CrossRef]

55. Braff, W.A.; Willner, D.; Hugenholtz, P.; Rabaey, K.; Buie, C.R. Dielectrophoresis-based discrimination of bacteria at the strain level based on their surface properties. *PLoS ONE* **2013**, *8*, e76751. [CrossRef]

56. Morgan, H.; Hughes, M.P.; Green, N.G. Separation of submicron bioparticles by dielectrophoresis. *Biophys. J.* **1999**, *77*, 516–525. [CrossRef]

57. Chou, C.F.; Tegenfeldt, J.O.; Bakajin, O.; Chan, S.S.; Cox, E.C.; Darnton, N.; Duke, T.; Austin, R.H. Electrodeless dielectrophoresis of single- and double-stranded DNA. *Biophys. J.* **2002**, *83*, 2170–2179. [CrossRef]

58. Jones, P.V.; Salmon, G.L.; Ros, A. Continuous separation of dna molecules by size using insulator-based dielectrophoresis. *Anal. Chem.* **2017**, *89*, 1531–1539. [CrossRef] [PubMed]

59. Liao, K.T.; Tsegaye, M.; Chaurey, V.; Chou, C.F.; Swami, N.S. Nano-constriction device for rapid protein preconcentration in physiological media through a balance of electrokinetic forces. *Electrophoresis* **2012**, *33*, 1958–1966. [CrossRef]

60. Abdallah, B.G.; Chao, T.C.; Kupitz, C.; Fromme, P.; Ros, A. Dielectrophoretic sorting of membrane protein nanocrystals. *ACS Nano* **2013**, *7*, 9129–9137. [CrossRef] [PubMed]

61. Lapizco-Encinas, B.H. On the recent developments of insulator-based dielectrophoresis: A review. *Electrophoresis* **2019**, *40*, 358–375. [CrossRef] [PubMed]

62. Price, J.A.; Burt, J.P.; Pethig, R. Applications of a new optical technique for measuring the dielectrophoretic behaviour of micro-organisms. *Biochim. Biophys. Acta* **1988**, *964*, 221–230. [CrossRef]

63. Pethig, R.; Huang, Y.; Wang, X.B.; Burt, J.P.H. Positive and negative dielectrophoretic collection of colloidal particles using interdigitated castellated microelectrodes. *J. Phys. D Appl. Phys.* **1992**, *25*, 881–888. [CrossRef]

64. Huang, Y.; Pethig, R. Electrode design for negative dielectrophoresis. *Meas. Sci. Technol.* **1991**, *2*, 1142–1146. [CrossRef]

65. Hoettges, K.F.; Hughes, M.P.; Cotton, A.; Hopkins, N.A.; McDonnell, M.B. Optimizing particle collection for enhanced surface-based biosensors. *IEEE Eng. Med. Biol. Mag.* **2003**, *22*, 68–74. [CrossRef]

66. Hoettges, K.F.; Hubner, Y.; Broche, L.M.; Ogin, S.L.; Kass, G.E.; Hughes, M.P. Dielectrophoresis-activated multiwell plate for label-free high-throughput drug assessment. *Anal. Chem.* **2008**, *80*, 2063–2068. [CrossRef]

67. Masuda, S.; Washizu, M.; Nanba, T. Novel method of cell fusion in field constriction area in fluid integration circuit. *IEEE Trans. Ind. Appl.* **1989**, *25*, 732–737. [CrossRef]

68. Pysher, M.D.; Hayes, M.A. Electrophoretic and dielectrophoretic field gradient technique for separating bioparticles. *Anal. Chem.* **2007**, *79*, 4552–4557. [CrossRef] [PubMed]

69. Braff, W.A.; Pignier, A.; Buie, C.R. High sensitivity three-dimensional insulator-based dielectrophoresis. *Lab Chip* **2012**, *12*, 1327–1331. [CrossRef] [PubMed]

70. Barrett, L.M.; Skulan, A.J.; Singh, A.K.; Cummings, E.B.; Fiechtner, G.J. Dielectrophoretic manipulation of particles and cells using insulating ridges in faceted prism microchannels. *Anal. Chem.* **2005**, *77*, 6798–6804. [CrossRef] [PubMed]

71. Hawkins, B.G.; Smith, A.E.; Syed, Y.A.; Kirby, B.J. Continuous-flow particle separation by 3D insulative dielectrophoresis using coherently shaped, dc-biased, ac electric fields. *Anal. Chem.* **2007**, *79*, 7291–7300. [CrossRef] [PubMed]

72. Zhu, J.; Xuan, X. Particle electrophoresis and dielectrophoresis in curved microchannels. *J. Colloid Interface Sci.* **2009**, *340*, 285–290. [CrossRef] [PubMed]

73. Cummings, E.B.; Singh, A.K. Dielectrophoretic trapping without embedded electrodes. In Proceedings of the SPIE: Conference on Microfluidic Devices and Systems III, Santa Clara, CA, USA, 18–19 September 2000; Volume 4177, pp. 164–173.

74. Cummings, E.B.; Singh, A.K. Dielectrophoresis in microchips containing arrays of insulating posts: Theoretical and experimental results. *Anal. Chem.* **2003**, *75*, 4724–4731. [CrossRef]

75. Camacho-Alanis, F.; Gan, L.; Ros, A. Transitioning streaming to trapping in DC insulator-based dielectrophoresis for biomolecules. *Sens. Actuators B Chem.* **2012**, *173*, 668–675. [CrossRef]

76. Hanasoge, S.; Devendra, R.; Diez, F.J.; Drazer, G. Electrokinetically driven deterministic lateral displacement for particle separation in microfluidic devices. *Microfluid. Nanofluidics* **2015**, *18*, 1195–1200. [CrossRef]

77. Beech, J.P.; Jonsson, P.; Tegenfeldt, J.O. Tipping the balance of deterministic lateral displacement devices using dielectrophoresis. *Lab Chip* **2009**, *9*, 2698–2706. [CrossRef]

78. Tran, T.S.H.; Ho, B.D.; Beech, J.P.; Tegenfeldt, J.O. Open channel deterministic lateral displacement for particle and cell sorting. *Lab Chip* **2017**, *17*, 3592–3600. [CrossRef]

79. Beech, J.P.; Keim, K.; Ho, B.D.; Guiducci, C.; Tegenfeldt, J.O. Active posts in deterministic lateral displacement devices. *Adv. Mater. Technol.* **2019**, *4*, 1900339. [CrossRef]

80. Calero, V.; Garcia-Sanchez, P.; Honrado, C.; Ramos, A.; Morgan, H. AC electrokinetic biased deterministic lateral displacement for tunable particle separation. *Lab Chip* **2019**, *19*, 1386–1396. [CrossRef] [PubMed]

81. Calero, V.; Garcia-Sanchez, P.; Ramos, A.; Morgan, H. Electrokinetic biased deterministic lateral displacement: Scaling analysis and simulations. *J. Chromatogr. A* **2020**, *1623*, 461151. [CrossRef] [PubMed]

82. Akbarzadeh, A.; Rezaei-Sadabady, R.; Davaran, S.; Joo, S.W.; Zarghami, N.; Hanifehpour, Y.; Samiei, M.; Kouhi, M.; Nejati-Koshki, K. Liposome: Classification, preparation, and applications. *Nanoscale Res. Lett.* **2013**, *8*, 102. [CrossRef]

83. Stremersch, S.; De Smedt, S.C.; Raemdonck, K. Therapeutic and diagnostic applications of extracellular vesicles. *J. Control. Release* **2016**, *244*, 167–183. [CrossRef] [PubMed]

84. Wiklander, O.P.B.; Brennan, M.A.; Lotvall, J.; Breakefield, X.O.; El Andaloussi, S. Advances in therapeutic applications of extracellular vesicles. *Sci. Transl. Med.* **2019**, *11*, eaav8521. [CrossRef]

85. Busatto, S.; Zendrini, A.; Radeghieri, A.; Paolini, L.; Romano, M.; Presta, M.; Bergese, P. The nanostructured secretome. *Biomater. Sci.* **2019**, *8*, 39–63. [CrossRef]

86. Kalluri, R.; LeBleu, V.S. The biology, function, and biomedical applications of exosomes. *Science* **2020**, *367*. [CrossRef]

87. Hochstetter, A.; Vernekar, R.; Austin, R.H.; Becker, H.; Beech, J.P.; Fedosov, D.A.; Gompper, G.; Kim, S.-C.; Smith, J.T.; Stolovitzky, G.; et al. Deterministic lateral displacement: Challenges and perspectives. *ACS Nano* **2020**, *14*, 10784–10795. [CrossRef]

88. Xia, Y.; Whitesides, G.M. Soft lithography. *Angew. Chem. Int. Ed. Engl.* **1998**, *37*, 550–575. [CrossRef]

89. Sbalzarini, I.F.; Koumoutsakos, P. Feature point tracking and trajectory analysis for video imaging in cell biology. *J. Struct. Biol.* **2005**, *151*, 182–195. [CrossRef] [PubMed]

90. Hunter, R.J.; Ottewill, R.H.; Rowell, R.L. *Zeta Potential in Colloid Science: Principles and Applications*; Academic Press: Cambridge, MA, USA, 1981.

91. Russel, W.B.; Russel, W.B.; Saville, D.A.; Schowalter, W.R. *Colloidal Dispersions*; Cambridge University Press: Cambridge, UK, 1991.

Micromachines **2020**, *11*, 1014

92. Bazant, M.Z. Nonlinear Electrokinetic Phenomena. In *Encyclopedia of Microfluidics and Nanofluidics*; Li, D., Ed.; Springer US: Boston, MA, USA, 2008; pp. 1461–1470. [CrossRef]

93. Green, N.G.; Ramos, A.; Gonzalez, A.; Morgan, H.; Castellanos, A. Fluid flow induced by nonuniform ac electric fields in electrolytes on microelectrodes. I. Experimental measurements. *Phys. Rev. E Stat. Phys. Plasmas Fluids Relat. Interdiscip. Top.* **2000**, *61*, 4011–4018. [CrossRef] [PubMed]

94. Gonzalez, A.; Ramos, A.; Green, N.G.; Castellanos, A.; Morgan, H. Fluid flow induced by nonuniform ac electric fields in electrolytes on microelectrodes. II. A linear double-layer analysis. *Phys. Rev. E Stat. Phys. Plasmas Fluids Relat. Interdiscip. Top.* **2000**, *61*, 4019–4028. [CrossRef] [PubMed]

95. Green, N.G.; Ramos, A.; Gonzalez, A.; Morgan, H.; Castellanos, A. Fluid flow induced by nonuniform ac electric fields in electrolytes on microelectrodes. III. Observation of streamlines and numerical simulation. *Phys. Rev. E Stat. Nonlin. Soft Matter. Phys.* **2002**, *66*, 026305. [CrossRef] [PubMed]

96. Ramos, A.; Morgan, H.; Green, N.G.; Castellanos, A. AC electric-field-induced fluid flow in microelectrodes. *J. Colloid Interface Sci.* **1999**, *217*, 420–422. [CrossRef]

97. Bazant, M.Z.; Squires, T.M. Induced-charge electrokinetic phenomena: Theory and microfluidic applications. *Phys. Rev. Lett.* **2004**, *92*, 066101. [CrossRef]

98. Levitan, J.A.; Devasenathipathy, S.; Studer, V.; Ben, Y.X.; Thorsen, T.; Squires, T.M.; Bazant, M.Z. Experimental observation of induced-charge electro-osmosis around a metal wire in a microchannel. *Colloids Surf. A-Physicochem. Eng. Asp.* **2005**, *267*, 122–132. [CrossRef]

99. Thamida, S.K.; Chang, H.C. Nonlinear electrokinetic ejection and entrainment due to polarization at nearly insulated wedges. *Phys. Fluids* **2002**, *14*, 4315–4328. [CrossRef]

100. Wang, Q.; Dingari, N.N.; Buie, C.R. Nonlinear electrokinetic effects in insulator-based dielectrophoretic systems. *Electrophoresis* **2017**, *38*, 2576–2586. [CrossRef]

101. Morgan, H.; Green, N.G. *AC Electrokinetics: Colloids and Nanoparticles*; Research Studies Press: Hertfordshire, UK, 2003.

102. Arnold, W.M.; Schwan, H.P.; Zimmermann, U. Surface conductance and other properties of latex-particles measured by electrorotation. *J. Phys. Chem.* **1987**, *91*, 5093–5098. [CrossRef]

103. Ermolina, I.; Morgan, H. The electrokinetic properties of latex particles: Comparison of electrophoresis and dielectrophoresis. *J. Colloid Interface Sci.* **2005**, *285*, 419–428. [CrossRef]

 *micromachines*

Review

Public-Health-Driven Microfluidic Technologies: From Separation to Detection

Xiangzhi Zhang [1,2], **Xiawei Xu** [2,3], **Jing Wang** [4], **Chengbo Wang** [4], **Yuying Yan** [5], **Aiguo Wu** [3] and **Yong Ren** [1,2,6,*]

1 Research Group for Fluids and Thermal Engineering, University of Nottingham Ningbo China, Ningbo 315100, China; xiangzhi.zhang2@nottingham.edu.cn
2 Department of Mechanical, Materials and Manufacturing Engineering, University of Nottingham Ningbo China, Ningbo 315100, China; xiawei.xu@nottingham.edu.cn
3 Cixi Institute of Biomedical Engineering, CAS Key Laboratory of Magnetic Materials and Devices & Key Laboratory of Additive Manufacturing Materials of Zhejiang Province, Ningbo Institute of Materials Technology and Engineering, Chinese Academy of Sciences, Ningbo 315201, China; aiguo@nimte.ac.cn
4 Department of Electrical and Electronic Engineering, University of Nottingham Ningbo China, Ningbo 315100, China; jing.wang@nottingham.edu.cn (J.W.); chengbo.wang@nottingham.edu.cn (C.W.)
5 Research Group for Fluids and Thermal Engineering, University of Nottingham, Nottingham NG7 2RD, UK; yuying.yan@nottingham.ac.uk
6 Key Laboratory of Carbonaceous Wastes Processing and Process Intensification Research of Zhejiang Province, University of Nottingham Ningbo China, Ningbo 315100, China
* Correspondence: yong.ren@nottingham.edu.cn

Abstract: Separation and detection are ubiquitous in our daily life and they are two of the most important steps toward practical biomedical diagnostics and industrial applications. A deep understanding of working principles and examples of separation and detection enables a plethora of applications from blood test and air/water quality monitoring to food safety and biosecurity; none of which are irrelevant to public health. Microfluidics can separate and detect various particles/aerosols as well as cells/viruses in a cost-effective and easy-to-operate manner. There are a number of papers reviewing microfluidic separation and detection, but to the best of our knowledge, the two topics are normally reviewed separately. In fact, these two themes are closely related with each other from the perspectives of public health: understanding separation or sorting technique will lead to the development of new detection methods, thereby providing new paths to guide the separation routes. Therefore, the purpose of this review paper is two-fold: reporting the latest developments in the application of microfluidics for separation and outlining the emerging research in microfluidic detection. The dominating microfluidics-based passive separation methods and detection methods are discussed, along with the future perspectives and challenges being discussed. Our work inspires novel development of separation and detection methods for the benefits of public health.

Keywords: microfluidic system; lab-on-a-chip; separation; detection; public health

Citation: Zhang, X.; Xu, X.; Wang, J.; Wang, C.; Yan, Y.; Wu, A.; Ren, Y. Public-Health-Driven Microfluidic Technologies: From Separation to Detection. *Micromachines* **2021**, *12*, 391. https://doi.org/10.3390/mi12040391

Academic Editor: Takasi Nisisako

Received: 28 February 2021
Accepted: 26 March 2021
Published: 2 April 2021

Publisher's Note: MDPI stays neutral with regard to jurisdictional claims in published maps and institutional affiliations.

1. Introduction

Public health is closely related to human wellbeing at diverse levels from our neighbor community to the national or even global security, covering the prevention, control, and treatment of major diseases, especially infectious diseases and noncommunicable chronic diseases, as well as supervision and control of food, drug, and public environmental sanitation. The infectious diseases include avian influenza, influenza, mad cow disease, an acquired immunodeficiency syndrome (AIDS), severe acute respiratory syndrome (SARS), and dengue fever, while noncommunicable chronic diseases include cancer, diabetes, and hypertension. When an infectious disease affects a large geographical area, it may cause death, destroy cities, politics, countries, disintegrate civilization, and even annihilate ethnic groups and species [1–4]. For example, the influenza pandemic claimed a high death toll in 1918, and SARS transmitted from bat broke out in 2002, affecting public health seriously [5,6]. Recently, SARS-CoV-2 virus has caused the unprecedented COVID-19 pandemic to occur and spread rapidly all over the world since December 2019 [7]. Up to 25 February 2021, there have been 111,999,954 confirmed cases and 2,486,679 confirmed death of COVID-19 around the world, posing a great threat to human health [8]. All these infectious diseases severely impact the development of the local economy and social stability. Infectious diseases can spread through air transmission, water transmission, food transmission, contact transmission, soil transmission, vertical transmission, body fluid transmission, and fecal oral transmission. Each infectious disease is caused by its specific pathogen, including viruses, bacteria, fungi, or parasites [9–12]. Based on the necessary conditions for infectious diseases such as the infection source, transmission route, and susceptible populations, three strategies can be deployed to manage infectious diseases via controlling the source of infection, cutting off the transmission routes, and isolating the susceptible populations, respectively. From the perspective of patients, the key lies in early detection, early diagnosis, early report, and early isolation. There are two main diagnostic targets for infectious diseases: pathogens or a specific antigen, antibody, or nucleic acid of an infectious pathogen [13–16]. Some of the techniques are time-consuming, labor-intensive, expensive, and unable to be carried out on-site detection because the use of bulky instruments is inevitable, which thereby hinders their applications and makes them insufficient to achieve rapid, accurate, and on-site diagnosis during a pandemic, especially in the most common and serious resource-poor areas [17,18].

In addition to infectious diseases, noncommunicable chronic diseases are also an important threat to human health, such as cardiovascular and cerebrovascular diseases, cancer, chronic respiratory diseases, and diabetes, which are mainly caused by unhealthy lifestyle and living environment. These kinds of diseases have a high incidence rate, disability rate, mortality rate, and medical expense, which can be thawed by early diagnosis and treatment. The common diagnostic methods in clinic for noncommunicable chronic diseases are tissue biopsy and liquid biopsy [19]. However, tissue biopsy is limited by sampling bias, sampling difficulty for deep tissue, and harm to patients, while liquid biopsy presents the challenges of a few samples, complex background, and gene typing polymorphism.

The health safety of food, drug, and public environmental sanitation has become a global question, such as excessive content of metals and additives, pesticide residues, and microbial contamination in food, water, gas, and soil. In the last few years, food safety accidents have occurred repeatedly [20,21]. Improved food safety analysis and testing are needed to control food contamination [22]. However, the traditional detection technology based on instrumental analysis has the disadvantages of expensive instruments, long cycle, large material consumption, complex operation, and low sensitivity, which cannot satisfy the demand of on-site, real-time, fast, and portable detection of food [23–25]. Meanwhile, with the increase in environment pollution, related detection, monitoring, and cleanup technologies should be developed to detect and collect toxic wastes and pollutions [26,27].

In the past decade, microfluidic technology has developed rapidly and microfluidics can lead to the combination of the sample pretreatment, separation, and detection processes into a small chip to realize the miniaturized, automated, and multifunction integrated analysis system, which find wide applications in molecular/cell biology, chemical/gene analysis, medicine, food safety, environment sensing, and other fields, because of the advantages such as less sample consumption, fast detection speed, facile operation, multifunctional integration, small size, and portability [28–30]. Among the numerous applications, microfluidic sensors have been developed to detect toxic gases in industrial wastewater, such as drinking water, heavy metals, and other waterborne pathogens. Microfluidic chip technology can be further integrated with electrochemical techniques, optical techniques, magnetic techniques, mass spectrometry, and other techniques to realize the separation and detection of targeted samples [31–33].

There are several reviews that focus on the application of microfluidic technologies in disease detection, food safety analysis, or environmental monitoring and detection. Nevertheless, there are inadequate studies focusing on unveiling the connection of microfluidics with public health, which has been arising as a global issue especially given the present COVID-19 crisis sweeping across the world. For example, it is helpful for determining infection risks to understand aerosol concentrations and persistence in public spaces because they play an important role in coronavirus transmission. However, it is difficult to measure the concentrations, which requires specialized equipment. The challenge may be tackled using microfluidics by taking advantaging of their high throughput capability and high integration level. Thanks to the advances in microfluidic development for cell separation and detection, point-of-care diagnostics are allowed, and monitoring of individual health conditions at home is possible, which greatly eases the public healthcare burden. The present study aims to give an overview of state-of-art microfluidic separation and detection technologies from the perspectives of public health, and we focus on separation and detection because they are two of the most important steps toward practical applications in disease detection, food safety analysis, and environmental monitoring and detection. The reviewed topics are closely associated with public health based on three aspects: (1) Prevention and early monitor of infectious diseases such as detection of COVID-19 viruses necessitates the demand to apply the microfluidics-based separation and detection methods; (2) Microfluidics also inspires novel routes to develop the vaccine products to effectively treat the diseases which may result in big public health impacts; and (3) The rapid growth of microfluidics-based separation and detection technologies also leads to point-of-care diagnosis which enables people to monitor their health conditions using portable devices at home, and this significantly mitigates the needs to seek medical assistance at hospitals and therefore promotes the public health level. The review paper is structured as follows: first, various microfluidic separation methods for public health are summarized and discussed. Subsequently, microfluidic detection methods applied to public health are systematically presented. Finally, the challenges and prospects of microfluidic separation and detection technology are discussed.

2. Microfluidic Separation Methods

Microfluidics technology is an interdisciplinary subject with many applications in various fields, such as biomedical, chemistry, disease diagnosis, and electronics industry [34]. Microfluidic devices have key functions in biomedical research, such as sample pretreatment, fluid processing, biosensing, separation and monitoring, and signal detection [35]. Among them, the microfluidic separation and classification of biological targets is quite essential for biological analysis and clinical diagnosis [36], which can be achieved with lab-on-a-chip (LOC), micrototal analytic systems (μTAS), and point-of-care (POC) diagnostics [37]. Although the development of microfluidic technology is still in its early stage, it has the potential to affect many fields from chemical synthesis and biological analysis to the disciplines of optics and information technology [38]. Microfluidic devices are able to create dynamic environments where the gradient of physiological conditions (such as pressure, temperature, and flow rate) can be kept constant, which have a low regent consumption and realize the quantitative assessment of cell migration [39]. For instance, the separation of cells to determine the content of biological molecules such as DNA, RNA, proteins, and lipids is essential in cell biology research, as well as diagnostic and therapeutic methods [40]. In the diagnosis of anemia, sorting and counting of red blood cells (RBCs) is of great importance [41]; in the diagnosis and treatment of HIV disease, the separation of CD4+ T cells from whole blood cells is essential [42]; and isolation of circulating tumor cells (CTCs) from blood cells is important for early diagnosis of cancer [43]. Microfluidic separation is also applied on the screening of cells, which is important in the detection of cancer cells [44,45]. The microfluidic separation of cells is based on their differences in physical properties [46]. When identifying CTCs, different cancer cells of epithelial origin need to be separated [47]. Suresh et al. investigated connections between single-cell mechanical properties and subcellular structural reorganization from biochemical factors in the context of gastrointestinal tumor and malaria [48]. It was found that cancer cells have larger sizes and higher deformability compared with healthy cells [49]. The deformability difference of normal red cells and red cells infected with malarial parasites can explain the mechanism of the spleen to remove parasitized red cells from the circulation of hosts [50]. In the past few decades, various separation and sorting methods have been developed for the separation of cells. Microfluidic separation and sorting has many advantages, including decreasing sample volumes, speeding up sample processing, enhancing sensitivity and spatial resolution, reducing device cost, increasing portability [51], reducing processing cost [52], raising efficiency [53], and contributing to environmental compatibility [54]. The application of polymer materials in microfluidic devices fabrication provides simple, cost-effective, and disposal advantages [55]. In order to avoid sample pollution by using biochemical markers, microfluidic techniques for label-free differentiation and fractional of cell population have been developed [40]. Droplets often act as microreactors for encapsulation. Since it can be important to ensure the droplets contain precise volume and composition or to ensure uniformity of emulsions, the separation and sorting of droplets should be taken in consideration, which can be realized by microfluidic approaches [56].

Microfluidics can be divided into two categories based on the scale: continuous microfluidics and digital microfluidics, [57–59]. Microparticle separation can be categorized as active and passive methods based on their manipulating forces [60]. In passive techniques, microfluidic devices do not use external forces for sorting or separation but rely purely on microfluidic phenomena and the interaction of the fluid with the geometrics of the microfluidic devices [61], while active sorting techniques involve an external field [62]. By comparing the advantages and disadvantages of passive and active techniques, Sajeesh and Kumar [36] concluded passive techniques are preferred in applications where energy input is of critical concern, whereas active separation techniques are preferred where higher particle sorting efficiency is required. The recent advances in separation and detection of whole-blood components were reviewed by Doddabasavana et al. [63]. The performance of microfluidic separation is evaluated according to the separation time, separation efficiency, throughput rate, and clogging filtration. According to separation approaches, separation techniques can be divided into passive and active methods [64]. The present paper focuses solely on passive separation/sorting approaches because they are easier to implement and thus can find more applications for public health, especially in developing countries or regions where people have limited access to costly apparatuses to energize the active separation approaches.

2.1. Pinched-Flow Fractionation (PFF)

The continuous sizing of particles in a microchannel is based on the characteristics of the laminar flow profile [65], and complicated outer field control is eliminated, which is usually required for other kinds of particle separation methods. Therefore, PFF can be applied both for particle analysis and for the preparation of monodispersed particles where energy input is of critical concern. The separation resolution in PFF is a function of the microchannel aspect ratio, particle size difference, and the microchannel sidewall roughness.

The work of Jain et al. [66] showed that particles with diameters on the order of the sidewall roughness cannot be separated in PFF devices with symmetric channels due to the same resistance in all outlet channels. Ma et al. [67] investigated the separation performance of an as PFF device by employing an immersed boundary-lattice Boltzmann method (IB-LBM), and the results showed that an adaptive regulating flux can be determined for each case to sort the cell mixture effectively. Yanai et al. [68] proposed a new hydrodynamic mechanism of particle separation in asPFF microchannel networks based on three-dimensional (3D) laminar flow profiles formed at intersections of lattice channels, and they confirmed that the depth of the main channel was critical for the particle separation efficiencies.

Berendsen et al. [69] proposed a microfluidic chip (Figure 1) based on the tumbling behavior of spermatozoa in pinched-flow fractionation which was used to separate spermatozoa from erythrocytes. Their study demonstrated a high extraction efficiency of 95% spermatozoa from a sample containing 2.5% spermatozoa while removing around 90% of the erythrocytes. Maenaka et al. [70] examined the availability of PFF for monodisperse droplets generated at the upstream T-junction via high-speed imaging. They reported a microfluidic system for continuous and size-dependent separation of droplets utilizing microscale hydrodynamics, which would be difficult for normal-scale schemes, such as centrifugation or filtration. Morijiri et al. [71] developed a microfluidic system based on the sedimentation effect of PFF, utilizing the inertial force of particle movement induced by the momentum change in the curved microchannel and the centrifugal force exerted on the flowing particles. In the study of Sai et al. [72], tunable pinched-flow fractionation (tunable PFF) was proposed as a modification of PFF with the introduction of a microvalve, where the effluent positions of the target particles can be controlled independently of the microchannel structure, which succeeded in separation micron and submicron-size polymer particles. Vig et al. [73] proposed a method for enhancing the separation of seven different polystyrene bead diameters ranging from 0.25 µm to 2.5 µm in PFF devices by a serpentine structure in the broadened segment, and the results demonstrated an amplification in the separation of up to 70%. Among the current microfluidic separation approaches, PFF is a cost-effective choice because of the simplicity of the device. However, there is a restriction for this method when vortices occur after the pinched segment with high Reynolds number (Re >> 1).

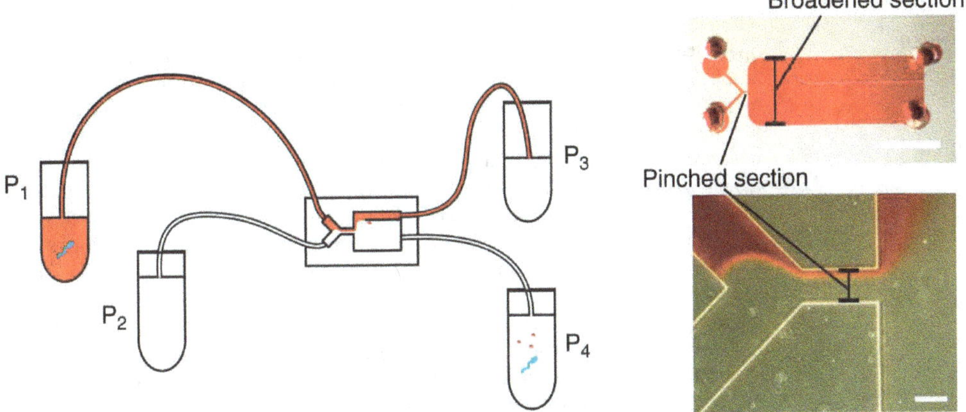

Figure 1. Pinched-flow fractionation (PFF) chip using a tumbling mechanism. The figure has been reproduced with permission from Springer Nature [69].

2.2. Inertia and Dean Flow

In fluid dynamics, secondary flow is a flow pattern, which is relatively weaker than the primary flow. The secondary flow can be controlled by the fluidic forces and the shape, size, and position of inserts [74]. In the study of macroscopic rigid spheres in Poiseuille flow by Segre and Silberberg particles migrated away from the wall and then accumulated at an equilibrium position of 0.6 from the axis around the tube radius due to lateral forces [75]. When a particle moves along a straight microchannel, two inertial lift forces are acting on the particle: shear-gradient-induced lift force, and wall-effect induced lift force [76]. Deformable particles contained in biomedical suspensions are underlying deformability-induced lift forces which lead to differences in dynamics [77]. The motion of a deformable particle in shear flow was studied by Bayareh and Mortazavi [78–80] with neglecting the gravity influence. Their results demonstrated that the equilibrium position of suspended particles is affected by the wall effect, deformability and sizes of particles, Reynolds number, density and viscosity ratio, etc. The nonlinear effects in finite-Reynolds-number flow were investigated, including the tubular pinch effect in cylindrical pipes [75]. Liu et al. [81] explored the focusing positions of different particle sizes in four focusing configurations for the separation of plasma, red blood cells, and cancer cells from the blood. The wall-induced inertia is significant in the thin layers near the walls where the lift is close to that calculated for linear shear flow, which increases dramatically with increasing Re above about 100 [82]. By analyzing the spatial distributions of spherical particles, Kim et al. [83] concluded the lateral migration of particles are induced by the high shear rate due to the small-scale effect and the particle equilibrium position as a function of Re. They observed the migration of particles markedly occurs at a very low Reynolds number and the critical Re when in the range of 20 to 30. Moreover, the inertial migration of spherical particles in a circular Poiseuille flow was numerically investigated with a Re smaller than 2200 [84]. A conclusion was drawn that the hydrodynamic interactions between the particles in different periodic cells have significant effects on the migration of the particles. The lateral migrations of viscous capsules [85], liquid drops, and vesicles [86] were also investigated.

Inertial microfluidics was applied in deformability-based cell classification and enrichment to reduce the complexity and costs of clinical applications [87]. Dean flow is a kind of secondary flow that can be generated by the fact that when a fluid flow in a curved pipe with a small radius of curvature, the flow has helical streamlines [88]. Focusing of particles suspended in solutions is largely independent of centrifugal forces, which suggests that Dean drag is the dominant lateral force to balance the influence of lift forces [89]. Di Carlo et al. [90] evaluated the migration attributed to lifting forces on particles in microfluidic devices by fabricating straight and curved microchannels under laminar flow conditions, when ordering is observed to be independent of particle buoyant direction. They developed a theoretical description of the underlying forces and a semiempirical relationship of cutoff and the channel geometry [91]. Inertia and Dean flow fractionation were applied in microfluidic separation and sorting of biochemical sample mixtures [40,75,92].

The concept of inertial microfluidics was used in continuous separation of a multiparticle mixture in a simple spiral microchannel coupled with rotational Dean drag [93]. In inertial microfluidic experiments, the particle diameters cannot be very small compared to the characteristic channel length scale, and the Reynolds number of the particle is in order of 10 [94]. A spiral lab-on-a-chip (LOC) was used for size-dependent focusing of particles at distinct equilibrium positions across the microchannel cross-section from a multiparticle mixture [95], which exhibited 90% separation efficiency. Lee at al. [96] developed a spiral microchannel system for the synchronization and selection of cancer cells at different phases of cell cycle of blood to predict the condition of disease as shown in Figure 2b. Yousuff et al. [97] proposed a new configuration of spiral channel, where collection outlets are a series of side-branching channels perpendicular to the main channel of egress in which closely spaced particle streams can be collected separately. A novel inertial separation technique using spiral microchannel having a stair-like cross-section was introduced for size-based particle separation [98]. A spiral microfluidic chip was also employed for continuous separation of CTCs [99] and sperm-like-particles (SLPs) [100] from blood.

The secondary flow induced by a microchannel with arc-shaped groove arrays was studied by Zhao et al. [101] with numerical approaches, and their results showed the secondary flow can guide different-size particles to the corresponding equilibrium positions. In the experiments, the performance of particles focusing was relatively insensitive to the variation of flow rate, which proves the availability of flow-insensitive microfluidic separation method in a reliable biosample preparation processing step for downstream bioassays. Yoon et al. [95] developed a size-selective separation system for microbeads by using secondary flow induced by centrifugal effects in a curved rectangular microchannel. The effects of curvature angles and channel heights on inertial focusing of microparticles in curvilinear microchannels were also investigated by Özbey et al. [102], and an optimum condition/configuration was obtained with a curvature angle of 280° at Re of 144 in the transition region.

Inertial size separation can be achieved in a contraction–expansion array (CEA) microchannel by a force balance between inertial lift and Dean drag forces in fluid regimes in which inertial fluid effects are significant [103]. In CEA systems, similar effects compared to Dean flows are produced by an abrupt change of the cross-sectional area, which is balanced by inertial lift forces throughout the contraction regions [81]. The CEA microchannels are applied for high-yield blood plasma separation with a level of 62.2% yield [104]. A fishbone-shaped microchannel was proposed by Kwak et al. [105] to separate platelets, erythrocytes, and leukocytes from human blood.

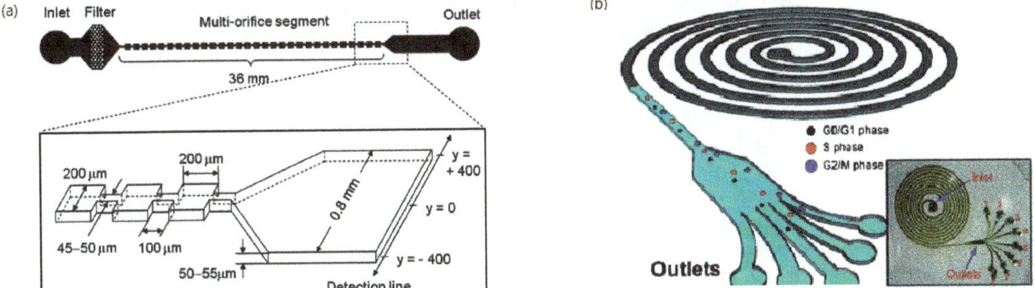

Figure 2. (**a**) Schematic diagram of multiorifice flow fractionation (MOFF) device. The figure has been reproduced with permission from the American Chemical Society [76]. (**b**) Schematic illustration of the spiral microfluidic design developed for cell-cycle synchronization. The figure has been reproduced with permission from the Royal Society of Chemistry [96].

Sim et al. [76] developed a novel separation method named as multiorifice flow fractionation (MOFF), where a microparticle moves laterally driven by the hydrodynamic inertial forces due to a multiorifice structure (Figure 2a). To improve the low efficiency of single-stage multiorifice flow fractionation (SS-MOFF) in separation for large particles, multistage multiorifice flow fractionation (MS-MOFF) was developed to isolate rare cells from human blood with a recovery increased from 73.2% to 88.7% while the purity slightly decreased from 91.4% to 89.1% [106]. A parallel multiorifice flow fractionation (p-MOFF) chip was developed and used for high-throughput size-based CTC separation, where CTCs can be focused at the center of the channel due to the wall-effect-induced lift force [107].

Separation of suspension in symmetric and asymmetric serpentine microchannels is also driven by inertial and Dean effects. Yuan et al. [108] investigated particle focusing under Dean flow coupled with elasto-inertial effects in symmetric serpentine microchannels, which demonstrated acceleration of particle focusing and reduction of channel length.

Compared with PFF, techniques based on inertia and Dean flow can be applied in higher Reynolds number flow since they are based on the balance of inertial shear-gradient-induced lift force and wall-effect-induced lift force, where the Reynolds number is generally in the range of 10–270 [34].

2.3. Deterministic Lateral Displacement (DLD)

Deterministic lateral displacement (DLD) is a microfluidic particle-separation device with asymmetric bifurcation of laminar flow around obstacles. When particles in solution moving through an array of obstacles, their paths are determined based on their sizes and deformability. The lateral displacement can be accumulated by a periodically arranged obstacle array which lead to a macroscopic change in migration angle, thus realizing particle separation [109]. Frechette et al. [110] used Stokesian dynamics simulation to study the dynamics of non-Brownian spheres suspended in a quiescent fluid and moving through a periodic array of solid obstacles under the action of a constant external force. It was found that moving particles were locked into periodic trajectories with an average orientation that coincides with one of the lattice directions. Generally, the arrangement of obstacle array has two configurations: a square array [111] and rhombic array obstacles (Figure 3a) [112]. The critical particle size for fractionation was investigated by Inglis et al. [113] who built a model based on the micropost geometry, where the fluid is driven by hydrodynamics or by electro-osmosis.

The fraction of whole-blood components and extraction of blood plasma without dilution was achieved by a continuous-flow deterministic array without dilution [114,115].

Blood components including white blood cells, red blood cells, and platelets can be separated by their hydrodynamic diameters from blood plasma at flow velocities of 1000 μm/s and volume rates up to 1 μm/min. A disposable parallel DLD device was applied for enrichment of leukocytes from blood with a throughput of greater than 1 mL/min [116]. With the utilization of an array of triangular instead of circular posts, the performance of DLD devices can be improved by reducing clogging, lowering hydrostatic pressure requirements, and increasing the range of displacement characteristics [117]. The DLD arrays with other shapes were investigated, including triangle [117], airfoil [118], I-shaped [119], L-shaped [120], asymmetric shape [121], and optimized shape [122], which have been used instead of cylindrical one. The elastomeric properties of PDMS were utilized to achieve tunable particle separation in DLD devices [123]. With the introduction of an external force, a concept of force-driven DLD was proposed [124]. For overdamped particles under the action of external forces, the trajectories are periodic, and the migration angle corresponds to a tangent bifurcation [125]. Devendra et al. [126] investigated the continuous size-based separation of suspended particles in gravity-driven deterministic lateral displacement (g-DLD) devices. (Figure 3b) In their experiments, directional locking angles were strongly depended on the size of the particle, and the results suggested that relatively small forcing angles are well suited for size-fractionation purposes. In an upscaled DLD device, larger gaps were utilized instead of micrometer-sized gaps between the posts, where particles above a critical size were better separated [127].

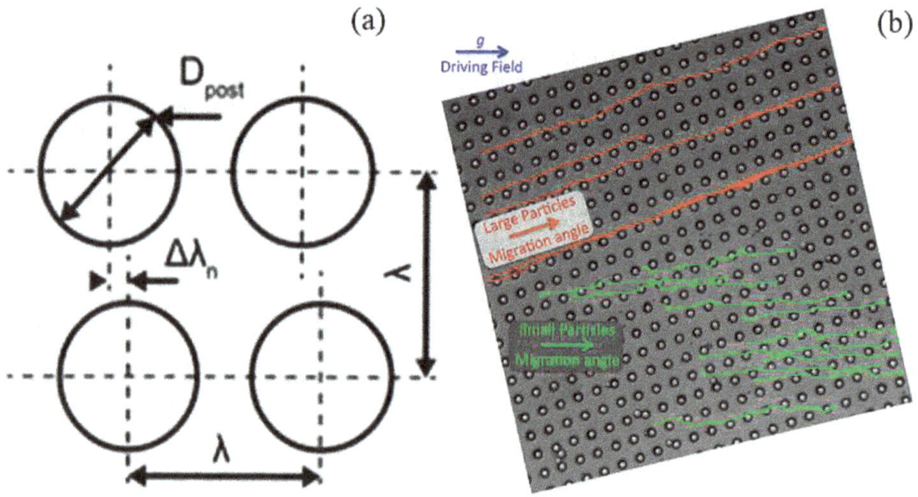

Figure 3. (**a**) Schematic of deterministic lateral displacement (DLD) chip with post placed at an angle to the flow direction. The figure has been reproduced with permission from the Royal Society of Chemistry [112]. (**b**) Microscopic image of gravity-driven deterministic lateral displacement (g-DLD) device. The figure has been reproduced with permission from the American Chemical Society [126].

DLD devices are also employed for the separation of CTCs [128], sleeping parasites [111], and deformable particles [129–131] by applying different pressures to the flowing fluids. A novel method for passive separation of microfluidic droplets by size using DLD was proposed by Joensson et al. [132], which showed a rate of 12,000 droplets/s with an 11 μm diameter. DLD separation for droplets can be accelerated by cell-induced shrinking [133]. A microfluidic DLD device was applied for spore purification to reduce the amount of debris in a suspension of fungal spores with almost 100% purity and recovery in continuously microspheres [134]. DLD techniques are suitable for the sorting of kinds of biological particles and droplets, but such a method requires an array of posts.

2.4. Microscale Filters

Microscale filters are widely employed in the separation of bioparticles/droplets based on size and deformability [40]. The most commonly used types of microfilters are categorized as dead-end mode [135], where the low is perpendicular to the filter structure, including membrane [136], planar [137], weir [138], pillar [139], and crossflow filters [114], where the flow is in the direction of the filter plane [140].

Membrane-based separation is a pressure-driven process [141,142], which has been widely used for microfiltration, ultrafiltration, reverse osmosis, ion-exchange, and gas separation [143]. The size-based crossflow separation can also be achieved using multistage arc-unit structures in a microfluidic device as shown in Figure 4 [141]. Chen et al. [144] proposed a method for preparation of microfiltration membranes made up with cellulose acetate (CA) blended with polyethyleneimine (PEI), where PEI can provide coupling sites for ligands in affinity separation or be used as a ligand for metal chelating, endotoxin removal, or ion exchange. In the study of Aussawasathien et al. [145], electrospun nylon-6 nanofibrous membranes were employed as prefilters for separation of micron to submicron particles from water due to their excellent chemical and thermal resistance as well as high wettability. A PDMS-membrane microfluidic immunosensor was used for rapid detection of foodborne pathogens integrated with a specific antibody-immobilized alumina nanoporous membrane. By sandwiching a filter membrane between a two-layer chip, Liu et al. [146] developed a vacuum-accelerated microfluidic immunoassay (VAMI), which could simultaneously achieve higher sensitivity and require less time compared with conventional microfluidic immunoassays. Nam et al. [136] proposed a novel effective manufacturing process that uses reusable 3D silicon molds with microneedle and microblade shapes to form submicron-sized nanopores and slit arrays in PDMS films. This process has been successfully applied to trap submicron-sized bacteria with a filter recovery rate of 90.1%. A superhydrophilic membrane with rough and hierarchical structures was used in the separation of oil-in-water emulsions since it can be fouled by surfactant-stabilized oil and organic foulants [147]. Ng et al. [148] designed and fabricated different gradient ceramic membranes including one-, two-, and three-layer ceramic membranes with a low total resistance, which demonstrated that the gradient porous membrane can be used to enhance the filtration performance.

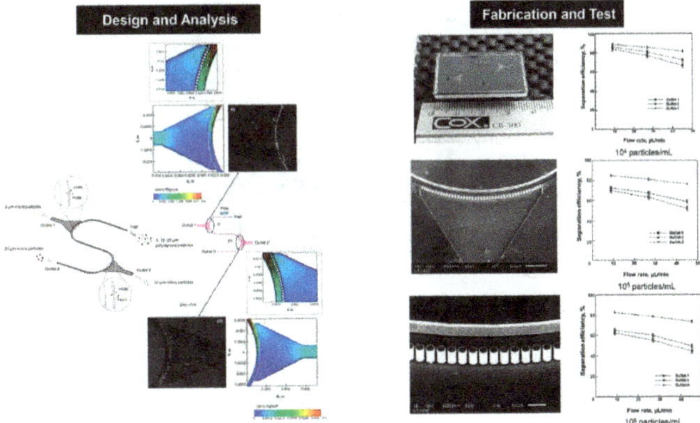

Figure 4. Crossflow microfilter with multistage dual arc-unit structures. The figure has been reproduced with permission from Elsevier [141].

Besides membranes, various types of microfabricated filters have been developed for microparticle separation. Crowley et al. [137] developed a planar microfilter for the isolation of plasma from whole blood with a separation efficiency three times higher than microporous membranes. An array of micropillars with a diameter of 12 μm and a height of 15 µm was arranged in I-shape as a filter for the separation of spherical and nonspherical particles [139]. Compared with pillar type, the microfilters of weir type show a higher separation efficiency due to the small gap of pillar [149]. A slanted weir microfluidic device was applied for the separation of CTCs from the peripheral blood, which showed a 97% separation efficiency as well as an 8-log depletion of erythrocytes and 5.6-log depletion of leukocytes [138]. As a modification, a cascading weir-type microfilter was constructed by Wu et al. [140] for plasma separation from blood samples.

The separation of microparticles was reported to be achieved in crossflow microfilters for cell biology research or various diagnostic and therapeutic applications, including cells extraction [150–152], plasma fabrication [137,153], leukapheresis [154], and myocytes/nonmyocytes from neonatal rat myocardium [155]. A microfluidic technique was proposed for separation of white blood cells (WBCs) from whole human blood, where the separation was performed in crossflow in an array of microchannels with a deep main channel and a large number of orthogonal and shallow side channels [151], as shown in Figure 5. The flow and shear stress characteristics inside a crossflow filter were studies by Mielink et al. [156] with employing microparticle image velocimetry (micro-PIV) measurements and computational fluid dynamics (CFD) analysis, demonstrating filter performance can be improved since substantial increase in the local wall shear can reduce clogging and cell cake formation.

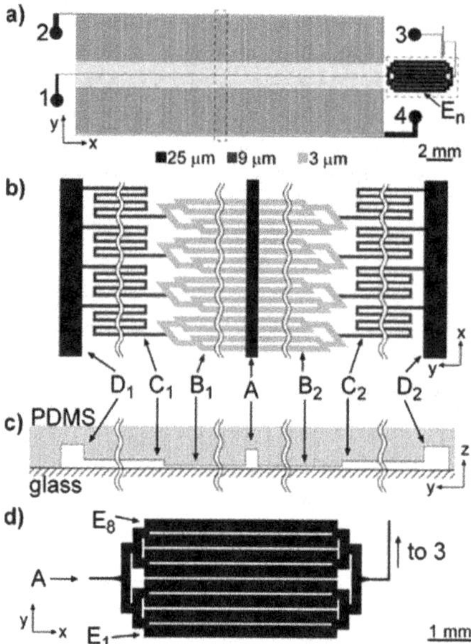

Figure 5. (a) Drawing of the microfluidic device, ports labeled 1–4 are blood inlet, perfusion inlet, WBC outlet, and RBC outlet, respectively; (b) blowup of a fragment of the separation network outlined with a dotted line in (a) turned counterclockwise by 90° with respect to (a); (c) cross-sectional view of channels in the separation network, dimensions are not to scale; (d) blowup of E channels outlined with a dotted line in (a). Channel depths, 25, 9, and 3 µm, are grayscale coded in (a,b,d). The figure has been reproduced with permission from the American Chemical Society [151].

Moorthy et al. [157] proposed in situ fabrication of porous filters using emulsion photopolymerization for microsystems to mimic the functionality of the centrifuge and power requirements as well as enabling the handling of small sample volumes. A novel microfluidic device constituted by microfilter, micromixer, micropillar array, microweir, microchannel, and microchamber was fabricated and used for isolation of WBCs from RBCs of whole blood [152]. Aran et al. [158] developed a microfiltration system consisted of a two-compartment mass exchanger with two aligned sets of PDMS microchannels, separated by a porous polycarbonate (PCTE) membrane. Lo et al. [159] described a multichamber device with porous membranes incorporated with variable pore sizes between the compartments within the microfluidic device, where nonhomogenous cell mixtures can be fractionated into different compartments in stages and collected for further analysis.

2.5. Other Hydrodynamic Methods

Besides the methods listed above, other hydrodynamic methods are also explored to be employed in separation of microparticles, including hydrodynamic filtration [160–163], Zweifach–Fung effect [164–169], trilobite separator [170–174], microvortex [175], and microhydrocyclone [176]. For particles flowing in a microchannel, their center positions cannot be at a certain position where the distance from sidewalls is equal to the particle radius. Yamada et al. [160] proposed the method of hydrodynamic filtration (HDF) for continuous concentration and classification of particles within microfluidic devices. By withdrawing a small amount of liquid repeatedly from the main streams through the side channels, particles are concentrated and arranged on the sidewalls by repeatedly drawing a small amount of liquid from the main flow through the side channel. Then, the concentrated and arranged particles can be collected through other side channels in downstream according to their sizes. Therefore, continuous introduction of the particle suspension into the microchannel can simultaneously perform particle concentration and classification. In this method, the flow profile inside the precisely manufactured microchannel determines the size limit of the filtered materials. Thus, the separation for small particles in much larger channels avoiding the problem of channel clogging. This device was applied for blood cell classification [161], as shown in Figure 6, and the sorting efficiency of hydrodynamic filtration device was dramatically improved by employing a flow splitting and recombination scheme [162]. Chiu et al. [163] proposed a microfluidic chip to separate microparticles using crossflow filtration enhanced with hydrodynamic focusing, which is needed to make soft lithograph fabrication to create microchannels and uses novel pressure bonding technology to make high-aspect-ratio filter structures.

Zweifach–Fung effect was the principle that a particle tends to follow the high-flow-rate channel when it reaches a bifurcation region [164]. This effect was employed for the separation of RBCs from plasma [165] and whole blood [166] and bacteria from blood [167]. The suspension stability of the blood was investigated by Fahraeus [168], and aggregation was observed to occur at a high concentration of blood under the influence of gravity and surface charge. Based on the characteristics of blood, Geng et al. [169] developed a device for separation of plasma from whole blood using a combination of Zweifach–Fung bifurcation law, centrifugation, and diffuser–nozzle effect.

Sample concentration or enrichment for rare particles in centrifugal separator often results in the cell being crushed and congregated during processing. Aiming to develop a nonclogging microconcentrator, Dong et al. [170] proposed a trilobite microchip for CaSki cells concentration using streamlined turbine blade-like micropillars based on the counter-flow principle. Hønsvall et al. [171] developed a microfluidic chip for continuously concentrating rigid cells in moving fluids based on a trilobite structure, which appears to be a promising tool for preconcentrating microalgae that are difficult to harvest due to their repelling properties or small size. The separation and concentration characteristics of the so-called trilobite separation unit was characterized experimentally by Mossige et al. [172]. With the introduction of a tunable structure, an increase in flow rate for low-pressure drops can be realized thus enabling clog-free particle separation of complex algal cells [173,174]. Besides the methods above, microvortex manipulator (MVM) [175] and microhydrocyclone [176] are also categorized as hydrodynamic methods for microfluidic separation and focusing of particles. The major public-health-related microfluidic separation/sorting technologies working in a passive way are summarized in Table 1.

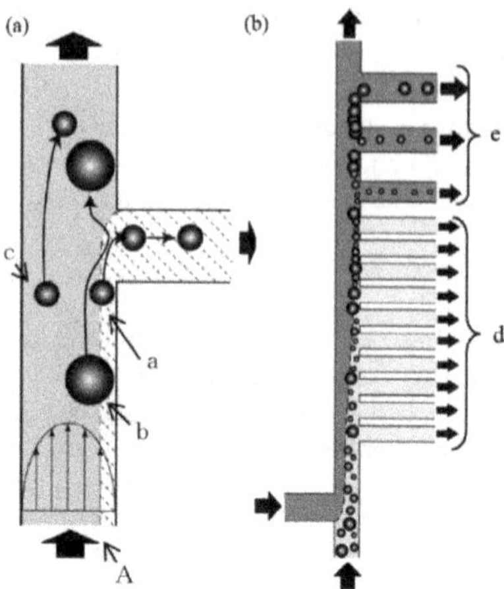

Figure 6. Principle of particle classification and concentration: (**a**) particle behavior at a branch point; (**b**) schematic diagram of particle classification and concentration in microchannel having multiple branch points and side channels. The figure has been reproduced with permission from John Wiley and Sons [161]. "a" represents a particle can enter the side channel, "b" and "c" represent particles that cannot enter the side channel; "d" represents the area where particles larger than a certain size cannot pass through; "e" represents the downstream area where particles are removed from the main stream in ascending order of size. A is the borderline and when a particle flows in the right region of the borderline, such a particle can enter the side channel.

Table 1. Public-health-related passive approaches for microfluidic separation.

Categories	Examples	References
Pinched-flow fractionation (PFF)	Symmetric PFF AsPFF Tumbling mechanism in PFF Sedimentation PFF Tunable PFF	[65] [67,68,70,73] [66] [71] [72]
Inertia and Dean flow	Inertial and Dean flow fractionation Spiral microchannel Curvature angles CEA Multiorifice Serpentine microchannel	[40,79,89,93] [97–100] [90,101,102] [83,103–105] [66,106,107] [108]
Deterministic lateral displacement (DLD)	DLD Disposable parallel DLD Optimized shape Tunable DLD Force-driven DLD Droplet shrinking Membrane	[109–115,127–131,134] [116] [117–122] [123] [124–126] [132,133] [136,144,145,147,148]
Microscale filter	Vacuum-accelerated microfluidic immunoassay (VAMI) Planar microfilter Weir microfluidic device Crossflow microfilter Porous filter Multicompartment	[146] [137,139] [138,140] [150–156] [156,158] [159]
Other hydrodynamic methods	Hydrodynamic filtration Zweifach–Fung effect Trilobite separator	[160–163] [164–169] [170–174]
	Microvortex	[175]
	Microhydrocyclone	[176]

3. Microfluidic Detection Methods

Microfluidic-method-integrated detection equipment has been becoming an ideal portable device for field sampling. Moreover, it improves the efficiency, sensitivity, and accuracy of detection and has advantages of rapid analysis, less usage of sample, and real-time characterization. Herein, microfluidic-based detection methods were summarized, including electrochemical detection, optical detection, and magnetic detection.

3.1. Electrochemical Detection

Electrochemical methods have advantages of shorter testing time, simpler device, and low cost, which can be classified into amperometric detection [177,178], impedimetric detection [179,180], and potentiometric detection [181]. Amperometric detection was formed when electroactive substances or electrolytes containing ions are under the action of an electric field, and they can be separated and detected effectively. Shiddiky et al. proposed an electrochemical detection method combined with micellar electrokinetic chromatography to separate and detect trace phenolic compounds in water [182]. They first used field-amplified sample stacking (FASS) and field-amplified sample injection (FASI) to separate the samples from water and then used cellulose-double-stranded DNA modified screen-printed carbon electrode to amplify the electrooxidation sensitivity of eight phenolic compounds. Hiraiwa et al. developed a method that used microtip immunoassay to detect the *Mycobacterium tuberculosis* (MTB) in sputum [183]. The microtip coated by antibodies was used to capture targeted bacteria. After that, the microtip surface would be covered by immunocomplex which can be detected by electric current. The detection limit of this method was 100 CFU per milliliter.

Impedimetric detection is a method using electrochemical impedance spectroscopy (EIS) for analysis. It has merits of the advantages of label-free and less amplitude disturbance [184]. As shown in Figure 7a, Cecchetto et al. proposed a label-free impedimetric detection method with a gold electrode modified by an anti-NS1 and a nonstructural dengue protein antibody to diagnose the dengue by detecting neat serum through the resistance changes resulting from the target binding [185].

Potentiometric detection is based on the potential change in an electrode in an electrochemical cell. The advantages of potentiometric biosensors are small volume, fast response, easy to use, low cost, anticolor, antiturbidity interference, and independent of sample volume [186,187]. For example, an electrochemical paper-based analytical device (EPAD) was designed to measure the concentrations of electrolyte ions ($Cl-$, $K+$, $Na+$, and $Ca2+$). In this design, ions were able to across the paper channels slowly so that accuracy was improved [188].

3.2. Optical Detection

Optical detection utilizes the properties of light, such as absorbance, fluorescence, and the emission mode of the sample when excited. Among optical detection methods, the fluorescence method is commonly used because it is sensitive, cheap, fast, and easy to operate [189]. The key to designing a fluorescence biosensor is fluorescent dyes or the labeling of fluorophores. Using fluorescence resonance energy transfer (FRET) is one of the most typical strategies, referring to the energy transfer from a donor fluorophore to an acceptor fluorophore [190]. Moreover, some nanomaterials also have fluorescence signals under specific conditions base on their unique properties of physical, chemical, and electronic transport. As shown in Figure 7b, Takemura et al. [191] designed an optical detection method using quantum-dots-based immunofluorescence to detect nonstructural protein 1 (NS1) of Zika virus. The fluorescence intensity signal was amplified and detected by a localized surface plasmon resonance (LSPR) signal from plasmonic gold nanoparticles (AuNPs). This sensor can detect NS1 of Zika virus ultrasensitively, rapidly, and quantitatively. In addition to the fluorescence method, absorbance of samples can be used to realize target analysis. For example, the analysis of UV absorption of nitrite samples can be used to determine the nitrite level in water [192].

Recently, surface-enhanced Raman scattering (SERS) spectroscopy has advantages of strong signal intensity, excellent photostability, biocompatibility, and especially the multiplexing ability, which makes it become a popular optical imaging and detection tool. For example, Wang et al. [193] first used folic acid (FA) functionalized gold (Au) SERS nanoparticles to detect CTCs in the presence of white blood cells successfully. Wu and co-workers have improved the sensitivity and specificity of CTC detection using the SERS properties of gold or silver with various shapes [194]. Moreover, Quang et al. [195] successfully demonstrated that the portable Raman spectrometer can be used to detect dipicolinic acid (DPA) and malachite green (MG) in real time, combined with a micropillar array chip.

3.3. Magnetic Detection

In the past few decades, the magnetic phenomenon of magnetic materials has been widely concerned, which is used to realize the sensitive detection of analytes [196]. Compared to the optical detection method, the magnetic detection method has advantages of low cost and high detection efficiency because of the elimination of expensive optical elements and the use of a magnetic field to shorten the sample preparation time [197,198]. Moreover, because biological samples have few magnetic background signals which can be ignored, the magnetic detection method has high specificity, sensitivity, and signal-to-noise ratio [199]. Hong et al. constructed an automated detection device for H7N9 influenza virus hemagglutinin, assisted by three-dimensional (3-D) magnetophoretic separation and magnetic label [200]. As shown in Figure 7c, a 3-D microchannel network with two-level channels was generated with multilayer glass slides under a magnetic field perpendicular to the microchannel. After the immunomagnetic separation, a magnetic-tagged complex was captured by an antibody-modified glass capillary, which causes the change of voltage in the miniature tube liquid sensor and therefore to obtain the detection signal. This work achieved the detection limit of 8.4 ng mL^{-1} for H$_7$N$_9$ hemagglutinin, with good specificity and reproducibility. Wu et al. [201] reported a Z-Lab point-of-care (POC) device which can detect swine influenza viruses sensitively and specifically reducing the dependence on the demands of sample treatment and operational skills sample handling and laboratory skill requirements. In this work, a portable and quantitative, giant magnetoresistive (GMR)-based immunoassay platform was designed to detect IAV nucleoprotein (NP) and purified H$_3$N$_2$v. It can achieve quantitative results within 10 min with a detection limitation of 15 ng per milliliter for IAV nucleoprotein, and 125 TCID50 per milliliter for purified H$_3$N$_2$v. Wu et al. [202] also introduced a new magnetic particle spectroscopy (MPS)-based biosensing scheme, where self-assembly magnetic nanoparticles (MNPs) can be used to detect H$_1$N$_1$ nucleoprotein molecules quantitatively. This work verified that it is reliable to use MPS and the self-assembly of MNPs to detect ultralow concentrations of targeted biomolecules, which can be applied on rapid, sensitive, and wash-free magnetic immunoassays.

Although these detection methods have good performance, they still have many shortcomings [203]. For electrochemical detection methods, they have high sensitivity, fast response, and low cost, but stability and susceptibility to interference are weak [204]. Optical detection methods have advantages of rapid response, flexibility, and experimental simplicity, but they are impacted by a high fluorescence background and short fluorescence lifetime [205]. Magnetic detection methods have advantages of low cost, high detection efficiency, high specificity, sensitivity, and signal-to-noise ratio but are limited by a shortage of miniaturized magnetic readout systems [206].

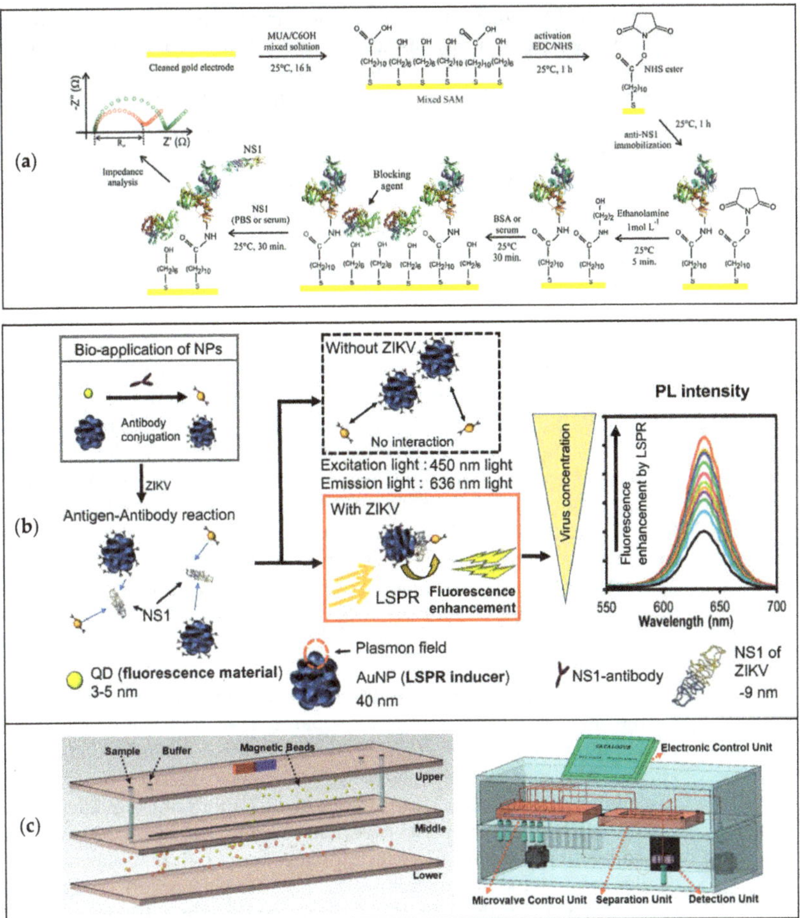

Figure 7. (**a**) Steps of electrode functionalization of the impedimetric biosensor to test neat serum for dengue diagnosis. The figure has been reproduced with permission from Elsevier [185]. (**b**) Schematic representation of the localized surface plasmon resonance (LSPR)-amplified immunofluorescence biosensor. The figure has been reproduced with permission from Takemura et al. [191]. (**c**) Schematic of the detection device based on the 3-D magnetophoretic separation and magnetic label. The figure has been reproduced with permission from the American Chemical Society [200].

4. Prospects of Microfluidics for Public Health Applications

In this paper, the emerging microfluidics studies for separation and detection have been overviewed, which have been widely applied in public health. In the context of an epidemic of infectious diseases, point-of-care diagnostics have become a matter of great concern, which enable people to implement home quarantine and real-time health monitoring. This method can cut off the source of infection and thus greatly reduce the rate of infection rate. Meanwhile, the fast development of microfluidics in the field of medicine enables point-of-care diagnostics to be realized. As mentioned above, microfluidics has advantages of less sample consumption, fast detection speed, facile operation, multifunctional integration, lower cost, and portability. The employment of microfluidic devices combined with point-of-care diagnostics can reduce the cost of public health care. Microfluidics can be well applied on virus detection, for example COVID-19 diagnosis. COVID-19 can be detected from saliva and respiratory samples of nasopharyngeal and oropharyngeal swabs by quantitative reverse-transcription polymerase chain reaction (qRT-PCR). COVID-19 can be identified through the variations of many biomarkers such as immunoglobulins, cytokines, and nucleic acids. Fast and accurate detection of these biomarkers by microfluidic system can be helpful in early diagnosis of COVID-19. Moreover, a microfluidic system combined with smartphones may realize the real-time health monitoring of individuals or populations during and after COVID-19 outbreaks. However, some detections such as impedance-based microfluidic devices and optical microfluidic devices require bulky instrumentation for the quantification of results. Besides, most samples require multiple pretreatments before detection. Therefore, although microfluidics combined with point-of-care diagnostics have the potential to allow the rapid detection of COVID-19 or other diseases, there is still a gap to be bridged.

Microfluidic techniques can also be applied in continuous production of vaccines [207]. For instance, the range of technology platforms for COVID-19 vaccines includes nucleic acid (DNA and RNA), virus-like particle, peptide, viral vector (replicating and nonreplicating), recombinant protein, live attenuated virus, and inactivated virus approaches, where microfluidic approaches can be applied [208–210]. Microfluidic devices were employed for vaccine therapy and delivery, especially for the administration of nucleic-acid-based vaccines by employing the host cell's transcriptional and translational capability to produce the desired protein, since uniform microspheres of DNA/RNA with a very narrow size distribution can be produced precisely [211]. Compared with other kinds of vaccines, because the vaccines of DNA or RNA do not have a viral coating, there is no requirement to invoke antibody reactions in order to suppress vaccine efficiency. Moreover, such vaccines are safe and easy to produce, thus presenting the opportunity for combining the genetic information of various antigen epitopes and cytokines [212].

5. Conclusions

Dramatic growth in microfluidic and lab-on-a-chip technologies has paved a way for the development of appropriate separation and detection-based diagnostics with the goal of improving local and global public health and thereby has attracted considerable efforts and resources in the past decade. Access to effective and efficient separation and detection methods has become increasingly important especially during the pandemic period. However, there exist several key factors that affect the introduction, acceptability, and sustainability of these technologies for practical applications; one of the greater challenges in deploying microfluidic diagnostic systems on a larger scale and to a wider extent is how to bring the cost down closer to the cost of the most inexpensive of current tests. The second challenge is that the performance of these methods is not good enough and needs to be further improved. This can be achieved by using a multistep method, which may lead to higher particle or cell separation performance. At the same time, a multistep method requires complicated configuration and a higher level of automation and integration technology. In addition, the production capacity of microfluids is far from meeting the actual needs. By increasing the number of devices running in parallel or the number of separation or detection units in the same microfluidic system, it is inevitable to enlarge the microfluidic technology. Accuracy and repeatability are also very crucial, and it is expected that an automated apparatus should be used as much as possible without much intervention from human operators. More sustainable efforts are required in the future to apply microfluidic technologies in developing more effective clinical or point-of-care tools, as well as detection systems to monitor the environmental conditions.

Author Contributions: Y.R. outlined the structure of the paper, X.Z., X.X. and Y.R. wrote the paper, J.W., C.W., A.W. and Y.Y. revised the paper. All authors have read and agreed to the published version of the manuscript.

Funding: This research was funded by National Natural Science Foundation of China, grant number NSFC31971292, Zhejiang Provincial Natural Science Foundation of China, grant number LY19E060001 and LQ19F050003, Zhejiang Provincial Department of Science and Technology, grant number 2020E10018, and Ningbo Science and Technology Bureau, grant number 2019F1030.

Acknowledgments: This work was financially supported by National Natural Science Foundation of China under Grant No. NSFC31971292, Zhejiang Provincial Natural Science Foundation of China under Grant No. LY19E060001 and LQ19F050003, Ningbo Science and Technology Bureau under Service Industry Science and Technology Programme with project code 2019F1030. The Zhejiang Provincial Department of Science and Technology is acknowledged for this research under its Provincial Key Laboratory Programme (2020E10018).

Conflicts of Interest: The authors declare no conflict of interest.

References

1. Cowie, B.C.; Dore, G.J. The perpetual challenge of infectious diseases. *N. Engl. J. Med.* **2012**, *367*, 89.
2. Daszak, P.; Cunningham, A.A.; Hyatt, A.D. Emerging infectious diseases of wildlife—Threats to biodiversity and human health. *Science* **2000**, *287*, 443–449. [CrossRef]
3. Johnson, P.T.J.; De Roode, J.C.; Fenton, A. Why infectious disease research needs community ecology. *Science* **2015**, *349*, 1259504. [CrossRef] [PubMed]
4. Jones, K.E.; Patel, N.G.; Levy, M.A.; Storeygard, A.; Balk, D.; Gittleman, J.L.; Daszak, P. Global trends in emerging infectious diseases. *Nature* **2008**, *451*, 990–993. [CrossRef] [PubMed]
5. Morens, D.M.; Fauci, A.S. The 1918 influenza pandemic: Insights for the 21st century. *J. Infect. Dis.* **2007**, *195*, 1018–1028. [CrossRef] [PubMed]
6. Ksiazek, T.G.; Erdman, D.; Goldsmith, C.S.; Zaki, S.R.; Peret, T.; Emery, S.; Tong, S.X.; Urbani, C.; Comer, J.A.; Lim, W.; et al. A novel coronavirus associated with severe acute respiratory syndrome. *N. Engl. J. Med.* **2003**, *348*, 1953–1966. [CrossRef] [PubMed]
7. Wu, Y.C.; Chen, C.S.; Chan, Y.J. The outbreak of COVID-19: An overview. *J. Chin. Med. Assoc.* **2020**, *83*, 217–220. [CrossRef] [PubMed]
8. World Health Organization. Geneva. 2021. Available online: https://www.who.int/ (accessed on 28 February 2021).
9. Boucher, H.W.; Talbot, G.H.; Bradley, J.S.; Edwards, J.E.; Gilbert, D.; Rice, L.B.; Scheld, M.; Spellberg, B. Bad bugs, no drugs: No ESKAPE! An update from the Infectious Diseases Society of America. *Clin. Infect. Dis.* **2009**, *48*, 1–12. [CrossRef]
10. McDonald, L.C.; Gerding, D.N.; Johnson, S.; Bakken, J.S.; Carroll, K.C.; Coffin, S.E.; Dubberke, E.R.; Garey, K.W.; Gould, C.V.; Kelly, C.; et al. Clinical practice guidelines for Clostridium difficile infection in adults and children: 2017 update by the Infectious Diseases Society of America (IDSA) and Society for Healthcare Epidemiology of America (SHEA). *Clin. Infect. Dis.* **2018**, *66*, e1–e48. [CrossRef]
11. Zumla, A.; Rao, M.; Wallis, R.S.; Kaufmann, S.H.E.; Rustomjee, R.; Mwaba, P.; Vilaplana, C.; Yeboah-Manu, D.; Chakaya, J.; Ippolito, G.; et al. Host-directed therapies for infectious diseases: Current status, recent progress, and future prospects. *Lancet Infect. Dis.* **2016**, *16*, e47–e63. [CrossRef]
12. Libertucci, J.; Young, V.B. The role of the microbiota in infectious diseases. *Nat. Microbiol.* **2019**, *4*, 35–45. [CrossRef]
13. You, M.; Li, Z.; Feng, S.; Gao, B.; Yao, C.; Hu, J.; Xu, F. Ultrafast photonic PCR based on photothermal nanomaterials. *Trends Biotechnol.* **2020**, *38*, 637–649. [CrossRef]
14. Yang, B.; Kong, J.; Fang, X. Bandage-like wearable flexible microfluidic recombinase polymerase amplification sensor for the rapid visual detection of nucleic acids. *Talanta* **2019**, *204*, 685–692. [CrossRef] [PubMed]
15. Liu, Z.; Shang, C.; Ma, H.; You, M. An upconversion nanoparticle-based photostable FRET system for long-chain DNA sequence detection. *Nanotechnology* **2020**, *31*, 235501. [CrossRef] [PubMed]
16. Wang, T.; Liu, Y.; Sun, H.H.; Yin, B.C.; Ye, B.C. An RNA-guided Cas9 nickase-based method for universal isothermal DNA amplification. *Angew. Chem. Int. Ed.* **2019**, *58*, 5382–5386. [CrossRef] [PubMed]
17. Liu, R.; Han, H.; Liu, F.; Lv, Z.H.; Wu, K.L.; Liu, Y.L.; Feng, Y.; Zhu, C.L. Positive rate of RT-PCR detection of SARS-CoV-2 infection in 4880 cases from one hospital in Wuhan, China, from Jan to Feb 2020. *Clin. Chim. Acta* **2020**, *505*, 172–175. [CrossRef]
18. Santiago, G.A.; Vazquez, J.; Courtney, S.; Matias, K.Y.; Andersen, L.E.; Colon, C.; Butler, A.E.; Roulo, R.; Bowzard, J.; Villanueva, J.M.; et al. Performance of the trioplex real-time RT-PCR assay for detection of Zika, dengue, and chikungunya viruses. *Nat. Commun.* **2018**, *9*, 1391. [CrossRef]
19. Shen, Z.Y.; Wu, A.G.; Chen, X.Y. Current detection technologies for circulating tumor cells. *Chem. Soc. Rev.* **2017**, *46*, 2038–2056. [CrossRef]
20. Lam, H.M.; Remais, J.; Fung, M.C. Food supply and food safety issues in China. *Lancet* **2013**, *381*, 2044–2053. [CrossRef]
21. Kaptan, G.; Fischer, A.R.H.; Frewer, L.J. Extrapolating understanding of food risk perceptions to emerging food safety cases. *J. Risk Res.* **2018**, *21*, 996–1018. [CrossRef]
22. Chiocchetti, G.D.M.E.; Piedra, C.A.J.; Monedero, V.; Cabrera, M.Z.; Devesa, V. Use of lactic acid bacteria and yeasts to reduce exposure to chemical food contaminants and toxicity. *Crit. Rev. Food Sci. Nutr.* **2019**, *59*, 15341545. [CrossRef]
23. Wu, W.; Yu, C.; Wang, Q.; Zhao, F.; He, H.; Liu, C.; Yang, Q. Research advances of DNA aptasensors for foodborne pathogen detection. *Crit. Rev. Food Sci. Nutr.* **2019**, *60*, 1636763. [CrossRef]
24. Cristina, L.; Elena, A.; Davide, C.; Marzia, G.; Lucia, D.; Cristiano, G.; Marco, A.; Carlo, R.; Laura, C.; Gabriella, G.M. Validation of a mass spectrometry-based method for milk traces detection in baked food. *Food Chem.* **2016**, *199*, 119–127. [CrossRef]
25. Wang, Y.; Duncan, T.V. Nanoscale sensors for assuring the safety of food products. *Curr. Opin. Biotechnol.* **2017**, *44*, 74–86. [CrossRef]

26. Yogarajah, N.; Tsai, S.S.H. Detection of trace arsenic in drinking water: Challenges and opportunities for microfluidics. *Environ. Sci. Water Res. Technol.* **2015**, *1*, 426–447. [CrossRef]

27. Ohira, S.-I.; Toda, K. Micro gas analysis system for measurement of atmospheric hydrogen sulfide and sulfur dioxide. *Lab Chip* **2005**, *5*, 1374–1379. [CrossRef]

28. Montes, R.J.; Ladd, A.J.C.; Butler, J.E. Transverse migration and microfluidic concentration of DNA using Newtonian buffers. *Biomicrofluidics* **2019**, *13*, 044104. [CrossRef]

29. Tweedie, M.; Sun, D.; Ward, B.; Maguire, P.D. Long-term hydrolytically stable bond formation for future membrane-based deep ocean microfluidic chemical sensors. *Lab Chip* **2019**, *19*, 1287–1295. [CrossRef]

30. An, X.; Zuo, P.; Ye, B.C. A single cell droplet microfluidic system for quantitative determination of food-borne pathogens. *Talanta* **2020**, *209*, 120571. [CrossRef] [PubMed]

31. Citartan, M.; Tang, T.H. Recent developments of aptasensors expedient for point-of-care (POC) diagnostics. *Talanta* **2019**, *199*, 556–566. [CrossRef] [PubMed]

32. Nicolini, A.M.; McCracken, K.E.; Yoon, J.Y. Future developments in biosensors for field-ready Zika virus diagnostics. *J. Biol. Eng.* **2017**, *11*, 7. [CrossRef]

33. Tepeli, Y.; Ülkü, A. Electrochemical biosensors for influenza virus a detection: The potential of adaptation of these devices to POC systems. *Sens. Actuators B Chem.* **2018**, *254*, 377–384. [CrossRef]

34. Nguyen, N.-T.; Wereley, S.T.; Shaegh, S.A.M. *Fundamentals and Applications of Microfluidics*; Artech House: Nordwood, MA, USA, 2019; ISBN 1630813656.

35. Chin, C.D.; Laksanasopin, T.; Cheung, Y.K.; Steinmiller, D.; Linder, V.; Parsa, H.; Wang, J.; Moore, H.; Rouse, R.; Umviligihozo, G. Microfluidics-based diagnostics of infectious diseases in the developing world. *Nat. Med.* **2011**, *17*, 1015. [CrossRef]

36. Sajeesh, P.; Sen, A.K. Particle separation and sorting in microfluidic devices: A review. *Microfluid. Nanofluid.* **2014**, *17*, 1–52. [CrossRef]

37. Yan, S.; Tan, S.H.; Li, Y.; Tang, S.; Teo, A.J.T.; Zhang, J.; Zhao, Q.; Yuan, D.; Sluyter, R.; Nguyen, N.-T. A portable, hand-powered microfluidic device for sorting of biological particles. *Microfluid. Nanofluid.* **2018**, *22*, 8. [CrossRef]

38. Whitesides, G.M. The origins and the future of microfluidics. *Nature* **2006**, *442*, 368–373. [CrossRef]

39. Coluccio, M.L.; D'Attimo, M.A.; Cristiani, C.M.; Candeloro, P.; Parrotta, E.; Dattola, E.; Guzzi, F.; Cuda, G.; Lamanna, E.; Carbone, E. A passive microfluidic device for chemotaxis studies. *Micromachines* **2019**, *10*, 551. [CrossRef]

40. Gossett, D.R.; Weaver, W.M.; Mach, A.J.; Hur, S.C.; Tse, H.T.K.; Lee, W.; Amini, H.; Di Carlo, D. Label-free cell separation and sorting in microfluidic systems. *Anal. Bioanal. Chem.* **2010**, *397*, 3249–3267. [CrossRef] [PubMed]

41. Miglierina, R.; Le Coniat, M.; Gendron, M.; Berger, R. Diagnosis of Fanconi's anemia by flow cytometry. *Nouv. Rev. Fr. Hematol.* **1990**, *32*, 391–393. [PubMed]

42. Cheng, X.; Irimia, D.; Dixon, M.; Sekine, K.; Demirci, U.; Zamir, L.; Tompkins, R.G.; Rodriguez, W.; Toner, M. A microfluidic device for practical label-free CD4+ T cell counting of HIV-infected subjects. *Lab Chip* **2007**, *7*, 170–178. [CrossRef]

43. Nagrath, S.; Sequist, L.V.; Maheswaran, S.; Bell, D.W.; Irimia, D.; Ulkus, L.; Smith, M.R.; Kwak, E.L.; Digumarthy, S.; Muzikansky, A. Isolation of rare circulating tumour cells in cancer patients by microchip technology. *Nature* **2007**, *450*, 1235–1239. [CrossRef]

44. Situma, C.; Hashimoto, M.; Soper, S.A. Merging microfluidics with microarray-based bioassays. *Biomol. Eng.* **2006**, *23*, 213–231. [CrossRef]

45. Dong, Y.; Skelley, A.M.; Merdek, K.D.; Sprott, K.M.; Jiang, C.; Pierceall, W.E.; Lin, J.; Stocum, M.; Carney, W.P.; Smirnov, D.A. Microfluidics and circulating tumor cells. *J. Mol. Diagn.* **2013**, *15*, 149–157. [CrossRef] [PubMed]

46. VAziri, A.; GopinAth, A. Cell and biomolecular mechanics in silico. *Nat. Mater.* **2008**, *7*, 15–23. [CrossRef] [PubMed]

47. Alshareef, M.; Metrakos, N.; Juarez Perez, E.; Azer, F.; Yang, F.; Yang, X.; Wang, G. Separation of tumor cells with dielectrophoresis-based microfluidic chip. *Biomicrofluidics* **2013**, *7*, 11803. [CrossRef] [PubMed]

48. Suresh, S.; Spatz, J.; Mills, J.P.; Micoulet, A.; Dao, M.; Lim, C.T.; Beil, M.; Seufferlein, T. Connections between single-cell biomechanics and human disease states: Gastrointestinal cancer and malaria. *Acta Biomater.* **2005**, *1*, 15–30. [CrossRef]

49. Suresh, S. Biomechanics and biophysics of cancer cells. *Acta Biomater.* **2007**, *3*, 413–438. [CrossRef]

50. Cranston, H.A.; Boylan, C.W.; Carroll, G.L.; Sutera, S.P.; Gluzman, I.Y.; Krogstad, D.J. Plasmodium falciparum maturation abolishes physiologic red cell deformability. *Science* **1984**, *223*, 400–403. [CrossRef]

51. Bhagat, A.A.S.; Bow, H.; Hou, H.W.; Tan, S.J.; Han, J.; Lim, C.T. Microfluidics for cell separation. *Med. Biol. Eng. Comput.* **2010**, *48*, 999–1014. [CrossRef]

52. Fernandez, R.E.; Rohani, A.; Farmehini, V.; Swami, N.S. Microbial analysis in dielectrophoretic microfluidic systems. *Anal. Chim. Acta* **2017**, *966*, 11–33. [CrossRef]

53. Nguyen, N.-T.; Hejazian, M.; Ooi, C.H.; Kashaninejad, N. Recent advances and future perspectives on microfluidic liquid handling. *Micromachines* **2017**, *8*, 186. [CrossRef]

54. Samiei, E.; Tabrizian, M.; Hoorfar, M. A review of digital microfluidics as portable platforms for lab-on a-chip applications. *Lab Chip* **2016**, *16*, 2376–2396. [CrossRef]

55. Tsao, C.-W. Polymer microfluidics: Simple, low-cost fabrication process bridging academic lab research to commercialized production. *Micromachines* **2016**, *7*, 225. [CrossRef]

56. Link, D.R.; Grasland-Mongrain, E.; Duri, A.; Sarrazin, F.; Cheng, Z.; Cristobal, G.; Marquez, M.; Weitz, D.A. Electric control of droplets in microfluidic devices. *Angew. Chem. Int. Ed.* **2006**, *45*, 2556–2560. [CrossRef]

57. Shukla, V.; Ali, N.B.Z.; Hussin, F.A.; Zwolinski, M. On testing of MEDA based digital microfluidics biochips. In Proceedings of the Fifth Asia Symposium on Quality Electronic Design (ASQED 2013), Penang, Malaysia, 26–28 August 2013; IEEE: Piscataway, NJ, USA, 2013; pp. 60–65.

58. Nguyen, N.-T.; Shaegh, S.A.M.; Kashaninejad, N.; Phan, D.-T. Design, fabrication and characterization of drug delivery systems based on lab-on-a-chip technology. *Adv. Drug Deliv. Rev.* **2013**, *65*, 1403–1419. [CrossRef]

59. Kashaninejad, N.; Chan, W.K.; Nguyen, N.-T. Fluid mechanics of flow through rectangular hydrophobic microchannels. In Proceedings of the International Conference on Nanochannels, Microchannels, and Minichannels, Edmonton, AB, Canada, 19–22 June 2011; ASME: New York, NY, USA, 2011; Volume 44632, pp. 647–655.

60. Yan, S.; Zhang, J.; Yuan, D.; Li, W. Hybrid microfluidics combined with active and passive approaches for continuous cell separation. *Electrophoresis* **2017**, *38*, 238–249. [CrossRef]

61. Lenshof, A.; Laurell, T. Continuous separation of cells and particles in microfluidic systems. *Chem. Soc. Rev.* **2010**, *39*, 1203–1217. [CrossRef] [PubMed]

62. Dalili, A.; Samiei, E.; Hoorfar, M. A review of sorting, separation and isolation of cells and microbeads for biomedical applications: Microfluidic approaches. *Analyst* **2019**, *144*, 87–113. [CrossRef]

63. Doddabasavana, G.; PadmaPriya, K.; Nagabhushana, K. A review of recent advances in separation and detection of whole blood components. *World J. Sci. Technol.* **2012**, *2*, 5–9.

64. Bayareh, M. An updated review on particle separation in passive microfluidic devices. *Chem. Eng. Process. Intensif.* **2020**, *153*, 107984. [CrossRef]

65. Oakey, J.; Allely, J.; Marr, D.W.M. Laminar-flow-based separations at the microscale. *Biotechnol. Prog.* **2002**, *18*, 1439–1442. [CrossRef]

66. Jain, A.; Posner, J.D. Particle dispersion and separation resolution of pinched flow fractionation. *Anal. Chem.* **2008**, *80*, 1641–1648. [CrossRef]

67. Ma, J.-T.; Xu, Y.-Q.; Tang, X.-Y. A numerical simulation of cell separation by simplified asymmetric pinched flow fractionation. *Comput. Math. Methods Med.* **2016**, *2016*, 2564584. [CrossRef] [PubMed]

68. Yanai, T.; Ouchi, T.; Yamada, M.; Seki, M. Hydrodynamic microparticle separation mechanism using three-dimensional flow profiles in dual-depth and asymmetric lattice-shaped microchannel networks. *Micromachines* **2019**, *10*, 425. [CrossRef] [PubMed]

69. Berendsen, J.T.W.; Eijkel, J.C.T.; Wetzels, A.M.; Segerink, L.I. Separation of spermatozoa from erythrocytes using their tumbling mechanism in a pinch flow fractionation device. *Microsyst. Nanoeng.* **2019**, *5*, 1–7. [CrossRef] [PubMed]

70. Maenaka, H.; Yamada, M.; Yasuda, M.; Seki, M. Continuous and size-dependent sorting of emulsion droplets using hydrodynamics in pinched microchannels. *Langmuir* **2008**, *24*, 4405–4410. [CrossRef]

71. Morijiri, T.; Sunahiro, S.; Senaha, M.; Yamada, M.; Seki, M. Sedimentation pinched-flow fractionation for size-and density-based particle sorting in microchannels. *Microfluid. Nanofluid.* **2011**, *11*, 105–110. [CrossRef]

72. Sai, Y.; Yamada, M.; Yasuda, M.; Seki, M. Continuous separation of particles using a microfluidic device equipped with flow rate control valves. *J. Chromatogr. A* **2006**, *1127*, 214–220. [CrossRef]

73. Vig, A.L.; Kristensen, A. Separation enhancement in pinched flow fractionation. *Appl. Phys. Lett.* **2008**, *93*, 203507. [CrossRef]

74. Stoecklein, D.; Wu, C.-Y.; Owsley, K.; Xie, Y.; Di Carlo, D.; Ganapathysubramanian, B. Micropillar sequence designs for fundamental inertial flow transformations. *Lab Chip* **2014**, *14*, 4197–4204. [CrossRef]

75. Di Carlo, D.; Edd, J.F.; Humphry, K.J.; Stone, H.A.; Toner, M. Particle segregation and dynamics in confined flows. *Phys. Rev. Lett.* **2009**, *102*, 94503. [CrossRef]

76. Park, J.-S.; Jung, H.-I. Multiorifice flow fractionation: Continuous size-based separation of microspheres using a series of contraction/expansion microchannels. *Anal. Chem.* **2009**, *81*, 8280–8288. [CrossRef]

77. Schaaf, C.; Stark, H. Inertial migration and axial control of deformable capsules. *Soft Matter* **2017**, *13*, 3544–3555. [CrossRef] [PubMed]

78. Bayareh, M.; Mortazavi, S. Numerical simulation of the motion of a single drop in a shear flow at finite Reynolds numbers. *Iran. J. Sci. Technol. Trans. B Eng.* **2009**, *33*, 441–452.

79. Bayareh, M.; Mortazavi, S. Binary collision of drops in simple shear flow at finite Reynolds numbers: Geometry and viscosity ratio effects. *Adv. Eng. Softw.* **2011**, *42*, 604–611. [CrossRef]

80. Mortazavi, S.; Bayareh, M. Geometry effects on the interaction of two equal-sized drops in simple shear flow at finite Reynolds numbers. *WIT Trans. Eng. Sci.* **2009**, *63*, 379–388.

81. Liu, L.; Han, L.; Shi, X.; Tan, W.; Cao, W.; Zhu, G. Hydrodynamic separation by changing equilibrium positions in contraction–expansion array channels. *Microfluid. Nanofluid.* **2019**, *23*, 52. [CrossRef]

82. Zeng, L.; Balachandar, S.; Fischer, P. Wall-induced forces on a rigid sphere at finite Reynolds number. *J. Fluid Mech.* **2005**, *536*, 1–25. [CrossRef]

83. Kim, Y.W.; Yoo, J.Y. The lateral migration of neutrally-buoyant spheres transported through square microchannels. *J. Micromech. Microeng.* **2008**, *18*, 65015. [CrossRef]

84. Shao, X.; Yu, Z.; Sun, B. Inertial migration of spherical particles in circular Poiseuille flow at moderately high Reynolds numbers. *Phys. Fluids* **2008**, *20*, 103307. [CrossRef]

85. Doddi, S.K.; Bagchi, P. Lateral migration of a capsule in a plane Poiseuille flow in a channel. *Int. J. Multiph. Flow* **2008**, *34*, 966–986. [CrossRef]

86. Abkarian, M.; Viallat, A. Dynamics of vesicles in a wall-bounded shear flow. *Biophys. J.* **2005**, *89*, 1055–1066. [CrossRef]
87. Hur, S.C.; Henderson-MacLennan, N.K.; McCabe, E.R.B.; Di Carlo, D. Deformability-based cell classification and enrichment using inertial microfluidics. *Lab Chip* **2011**, *11*, 912–920. [CrossRef] [PubMed]
88. Yoon, D.H.; Ha, J.B.; Bahk, Y.K.; Arakawa, T.; Shoji, S.; Go, J.S. Size-selective separation of micro beads by utilizing secondary flow in a curved rectangular microchannel. *Lab Chip* **2009**, *9*, 87–90. [CrossRef]
89. Russom, A.; Gupta, A.K.; Nagrath, S.; Di Carlo, D.; Edd, J.F.; Toner, M. Differential inertial focusing of particles in curved low-aspect-ratio microchannels. *New J. Phys.* **2009**, *11*, 75025. [CrossRef]
90. Di Carlo, D.; Irimia, D.; Tompkins, R.G.; Toner, M. Continuous inertial focusing, ordering, and separation of particles in microchannels. *Proc. Natl. Acad. Sci. USA* **2007**, *104*, 18892–18897. [CrossRef]
91. Di Carlo, D.; Edd, J.F.; Irimia, D.; Tompkins, R.G.; Toner, M. Equilibrium separation and filtration of particles using differential inertial focusing. *Anal. Chem.* **2008**, *80*, 2204–2211. [CrossRef] [PubMed]
92. Pamme, N. Continuous flow separations in microfluidic devices. *Lab Chip* **2007**, *7*, 1644–1659. [CrossRef]
93. Chatterjee, A. Size-Dependant Separation of Multiple Particles in Spiral Microchannels. Ph.D. Thesis, University of Cincinnati, Cincinnati, OH, USA, 2011.
94. Hood, K.; Lee, S.; Roper, M. Inertial migration of a rigid sphere in three-dimensional Poiseuille flow. *J. Fluid Mech.* **2015**, *765*, 452–479. [CrossRef]
95. Bhagat, A.A.S.; Kuntaegowdanahalli, S.S.; Papautsky, I. Continuous particle separation in spiral microchannels using dean flows and differential migration. *Lab Chip* **2008**, *8*, 1906–1914. [CrossRef]
96. Lee, W.C.; Bhagat, A.A.S.; Huang, S.; Van Vliet, K.J.; Han, J.; Lim, C.T. High-throughput cell cycle synchronization using inertial forces in spiral microchannels. *Lab Chip* **2011**, *11*, 1359–1367. [CrossRef]
97. Mohamed Yousuff, C.; Hamid, N.H.B.; Kamal Basha, I.H.; Wei Ho, E.T. Output channel design for collecting closely-spaced particle streams from spiral inertial separation devices. *AIP Adv.* **2017**, *7*, 85004. [CrossRef]
98. Ghadami, S.; Kowsari-Esfahan, R.; Saidi, M.S.; Firoozbakhsh, K. Spiral microchannel with stair-like cross section for size-based particle separation. *Microfluid. Nanofluid.* **2017**, *21*, 115. [CrossRef]
99. Chen, H. A triplet parallelizing spiral microfluidic chip for continuous separation of tumor cells. *Sci. Rep.* **2018**, *8*, 1–8. [CrossRef]
100. Son, J.; Jafek, A.R.; Carrell, D.T.; Hotaling, J.M.; Gale, B.K. Sperm-like-particle (SLP) behavior in curved microfluidic channels. *Microfluid. Nanofluid.* **2019**, *23*, 4. [CrossRef]
101. Zhao, Q.; Yuan, D.; Yan, S.; Zhang, J.; Du, H.; Alici, G.; Li, W. Flow rate-insensitive microparticle separation and filtration using a microchannel with arc-shaped groove arrays. *Microfluid. Nanofluid.* **2017**, *21*, 55. [CrossRef]
102. Özbey, A.; Karimzadehkhouei, M.; Bayrak, Ö.; Koşar, A. Inertial focusing of microparticles in curvilinear microchannels with different curvature angles. *Microfluid. Nanofluid.* **2018**, *22*, 62. [CrossRef]
103. Lee, M.G.; Choi, S.; Park, J.-K. Inertial separation in a contraction-expansion array microchannel. *J. Chromatogr. A* **2011**, *1218*, 4138–4143. [CrossRef]
104. Lee, M.G.; Choi, S.; Kim, H.-J.; Lim, H.K.; Kim, J.-H.; Huh, N.; Park, J.-K. High-yield blood plasma separation by modulating inertial migration in a contraction-expansion array microchannel. In Proceedings of the 16th International Solid-State Sensors, Actuators and Microsystems Conference, Beijing, China, 5–9 June 2011; IEEE: Piscataway, NJ, USA, 2011; pp. 258–261.
105. Kwak, B.; Lee, S.; Lee, J.; Lee, J.; Cho, J.; Woo, H.; Heo, Y.S. Hydrodynamic blood cell separation using fishbone shaped microchannel for circulating tumor cells enrichment. *Sens. Actuators B Chem.* **2018**, *261*, 38–43. [CrossRef]
106. Kwon, K.; Sim, T.; Moon, H.-S.; Lee, J.-G.; Park, J.C.; Jung, H.-I. A novel particle separation method using multi-stage multi-orifice flow fractionation (MS-MOFF). In Proceedings of the 14th International Conference on Miniaturized Systems for Chemistry and Life Sciences, Groningen, The Netherland, 3–7 October 2010.
107. Hyun, K.-A.; Koo, G.-B.; Han, H.; Sohn, J.; Choi, W.; Kim, S.-I.; Jung, H.-I.; Kim, Y.-S. Epithelial-to-mesenchymal transition leads to loss of EpCAM and different physical properties in circulating tumor cells from metastatic breast cancer. *Oncotarget* **2016**, *7*, 24677. [CrossRef]
108. Yuan, D.; Sluyter, R.; Zhao, Q.; Tang, S.; Yan, S.; Yun, G.; Li, M.; Zhang, J.; Li, W. Dean-flow-coupled elasto-inertial particle and cell focusing in symmetric serpentine microchannels. *Microfluid. Nanofluid.* **2019**, *23*, 41. [CrossRef]
109. Balvin, M.; Sohn, E.; Iracki, T.; Drazer, G.; Frechette, J. Directional locking and the role of irreversible interactions in deterministic hydrodynamics separations in microfluidic devices. *Phys. Rev. Lett.* **2009**, *103*, 78301. [CrossRef]
110. Frechette, J.; Drazer, G. Directional locking and deterministic separation in periodic arrays. *J. Fluid Mech.* **2009**, *627*, 379. [CrossRef]
111. Holm, S.H.; Beech, J.P.; Barrett, M.P.; Tegenfeldt, J.O. Separation of parasites from human blood using deterministic lateral displacement. *Lab Chip* **2011**, *11*, 1326–1332. [CrossRef]
112. Long, B.R.; Heller, M.; Beech, J.P.; Linke, H.; Bruus, H.; Tegenfeldt, J.O. Multidirectional sorting modes in deterministic lateral displacement devices. *Phys. Rev. E* **2008**, *78*, 46304. [CrossRef]
113. Inglis, D.W.; Davis, J.A.; Austin, R.H.; Sturm, J.C. Critical particle size for fractionation by deterministic lateral displacement. *Lab Chip* **2006**, *6*, 655–658. [CrossRef]
114. Davis, J.A.; Inglis, D.W.; Morton, K.J.; Lawrence, D.A.; Huang, L.R.; Chou, S.Y.; Sturm, J.C.; Austin, R.H. Deterministic hydrodynamics: Taking blood apart. *Proc. Natl. Acad. Sci. USA* **2006**, *103*, 14779–14784. [CrossRef] [PubMed]
115. Beech, J.P.; Holm, S.H.; Adolfsson, K.; Tegenfeldt, J.O. Sorting cells by size, shape and deformability. *Lab Chip* **2012**, *12*, 1048–1051. [CrossRef] [PubMed]

116. Inglis, D.W.; Lord, M.; Nordon, R.E. Scaling deterministic lateral displacement arrays for high throughput and dilution-free enrichment of leukocytes. *J. Micromech. Microeng.* **2011**, *21*, 54024. [CrossRef]
117. Loutherback, K.; Chou, K.S.; Newman, J.; Puchalla, J.; Austin, R.H.; Sturm, J.C. Improved performance of deterministic lateral displacement arrays with triangular posts. *Microfluid. Nanofluid.* **2010**, *9*, 1143–1149. [CrossRef]
118. Dincau, B.M.; Aghilinejad, A.; Chen, X.; Moon, S.Y.; Kim, J.-H. Vortex-free high-Reynolds deterministic lateral displacement (DLD) via airfoil pillars. *Microfluid. Nanofluid.* **2018**, *22*, 137. [CrossRef]
119. Zeming, K.K.; Ranjan, S.; Zhang, Y. Rotational separation of non-spherical bioparticles using I-shaped pillar arrays in a microfluidic device. *Nat. Commun.* **2013**, *4*, 1–8. [CrossRef]
120. Ranjan, S.; Zeming, K.K.; Jureen, R.; Fisher, D.; Zhang, Y. DLD pillar shape design for efficient separation of spherical and non-spherical bioparticles. *Lab Chip* **2014**, *14*, 4250–4262. [CrossRef] [PubMed]
121. Au, S.H.; Edd, J.; Stoddard, A.E.; Wong, K.H.K.; Fachin, F.; Maheswaran, S.; Haber, D.A.; Stott, S.L.; Kapur, R.; Toner, M. Microfluidic isolation of circulating tumor cell clusters by size and asymmetry. *Sci. Rep.* **2017**, *7*, 1–10. [CrossRef] [PubMed]
122. Hyun, J.; Hyun, J.; Wang, S.; Yang, S. Improved pillar shape for deterministic lateral displacement separation method to maintain separation efficiency over a long period of time. *Sep. Purif. Technol.* **2017**, *172*, 258–267. [CrossRef]
123. Beech, J.P.; Tegenfeldt, J.O. Tuneable separation in elastomeric microfluidics devices. *Lab Chip* **2008**, *8*, 657–659. [CrossRef]
124. Bowman, T.; Frechette, J.; Drazer, G. Force driven separation of drops by deterministic lateral displacement. *Lab Chip* **2012**, *12*, 2903–2908. [CrossRef]
125. Herrmann, J.; Karweit, M.; Drazer, G. Separation of suspended particles in microfluidic systems by directional locking in periodic fields. *Phys. Rev. E* **2009**, *79*, 61404. [CrossRef]
126. Devendra, R.; Drazer, G. Gravity driven deterministic lateral displacement for particle separation in microfluidic devices. *Anal. Chem.* **2012**, *84*, 10621–10627. [CrossRef]
127. Lubbersen, Y.S.; Dijkshoorn, J.P.; Schutyser, M.A.I.; Boom, R.M. Visualization of inertial flow in deterministic ratchets. *Sep. Purif. Technol.* **2013**, *109*, 33–39. [CrossRef]
128. Loutherback, K.; D'Silva, J.; Liu, L.; Wu, A.; Austin, R.H.; Sturm, J.C. Deterministic separation of cancer cells from blood at 10 mL/min. *AIP Adv.* **2012**, *2*, 42107. [CrossRef] [PubMed]
129. Zhang, Z.; Henry, E.; Gompper, G.; Fedosov, D.A. Behavior of rigid and deformable particles in deterministic lateral displacement devices with different post shapes. *J. Chem. Phys.* **2015**, *143*, 243145. [CrossRef]
130. Quek, R.; Le, D.V.; Chiam, K.-H. Separation of deformable particles in deterministic lateral displacement devices. *Phys. Rev. E* **2011**, *83*, 56301. [CrossRef]
131. Ghasemi, M.; Holm, S.H.; Beech, J.P.; Björnmalm, M.; Tegenfeldt, J.O. Separation of deformable hydrogel microparticles in deterministic lateral displacement devices. In Proceedings of the 16th International Conference on Miniaturized Systems for Chemistry and Life Sciences, MicroTAS 2012, Okinawa, Japan, 28 October–1 November 2012; Chemical and Biological Microsystems Society: Washington, DC, USA, 2012; pp. 1672–1674.
132. Joensson, H.N.; Uhlén, M.; Svahn, H.A. Deterministic lateral displacement device for droplet separation by size—Towards rapid clonal selection based on droplet shrinking. In Proceedings of the 14th International Conference on Miniaturized Systems for Chemistry and Life Sciences, Groningen, The Netherlands, 3–7 October 2010.
133. Joensson, H.N.; Uhlén, M.; Svahn, H.A. Droplet size based separation by deterministic lateral displacement—Separating droplets by cell-induced shrinking. *Lab Chip* **2011**, *11*, 1305–1310. [CrossRef]
134. Inglis, D.W.; Herman, N.; Vesey, G. Highly accurate deterministic lateral displacement device and its application to purification of fungal spores. *Biomicrofluidics* **2010**, *4*, 24109. [CrossRef] [PubMed]
135. Zheng, S.; Lin, H.; Liu, J.-Q.; Balic, M.; Datar, R.; Cote, R.J.; Tai, Y.-C. Membrane microfilter device for selective capture, electrolysis and genomic analysis of human circulating tumor cells. *J. Chromatogr. A* **2007**, *1162*, 154–161. [CrossRef]
136. Nam, Y.-H.; Lee, S.-K.; Kim, J.-H.; Park, J.-H. PDMS membrane filter with nano-slit array fabricated using three-dimensional silicon mold for the concentration of particles with bacterial size range. *Microelectron. Eng.* **2019**, *215*, 111008. [CrossRef]
137. Crowley, T.A.; Pizziconi, V. Isolation of plasma from whole blood using planar microfilters for lab-on-a-chip applications. *Lab Chip* **2005**, *5*, 922–929. [CrossRef] [PubMed]
138. Yoon, Y.; Lee, J.; Ra, M.; Gwon, H.; Lee, S.; Kim, M.Y.; Yoo, K.-C.; Sul, O.; Kim, C.G.; Kim, W.-Y. Continuous separation of circulating tumor cells from whole blood using a slanted weir microfluidic device. *Cancers* **2019**, *11*, 200. [CrossRef] [PubMed]
139. Indhu, R.; Mercy, A.S.; Shreemathi, K.M.; Radha, S.; Kirubaveni, S.; Sreeja, B.S. Design of a Filter Using Array of Pillar for Particle Separation. *Mater. Today Proc.* **2018**, *5*, 10889–10894. [CrossRef]
140. Wu, C.-C.; Hong, L.-Z.; Ou, C.-T. Blood cell-free plasma separated from blood samples with a cascading weir-type microfilter using dead-end filtration. *J. Med. Biol. Eng.* **2012**, *32*, 163–168. [CrossRef]
141. Lee, Y.-T.; Dang, C.; Hong, S.; Yang, A.-S.; Su, T.-L.; Yang, Y.-C. Microfluidics with new multi-stage arc-unit structures for size-based cross-flow separation of microparticles. *Microelectron. Eng.* **2019**, *207*, 37–49. [CrossRef]
142. Strathmann, H. Membrane separation processes. *J. Memb. Sci.* **1981**, *9*, 121–189. [CrossRef]
143. Strathmann, H. Membrane separation processes: Current relevance and future opportunities. *AIChE J.* **2001**, *47*, 1077–1087. [CrossRef]
144. Chen, Z.; Deng, M.; Chen, Y.; He, G.; Wu, M.; Wang, J. Preparation and performance of cellulose acetate/polyethyleneimine blend microfiltration membranes and their applications. *J. Membr. Sci.* **2004**, *235*, 73–86. [CrossRef]

145. Aussawasathien, D.; Teerawattananon, C.; Vongachariya, A. Separation of micron to sub-micron particles from water: Electrospun nylon-6 nanofibrous membranes as pre-filters. *J. Membr. Sci.* **2008**, *315*, 11–19. [CrossRef]

146. Liu, Y.; Yu, J.; Du, M.; Wang, W.; Zhang, W.; Wang, Z.; Jiang, X. Accelerating microfluidic immunoassays on filter membranes by applying vacuum. *Biomed. Microdevices* **2012**, *14*, 17–23. [CrossRef]

147. Shao, S.; Liu, Y.; Shi, D.; Qing, W.; Fu, W.; Li, J.; Fang, Z.; Chen, Y. Control of organic and surfactant fouling using dynamic membranes in the separation of oil-in-water emulsions. *J. Colloid Interface Sci.* **2020**, *560*, 787–794. [CrossRef]

148. Ng, T.C.A.; Lyu, Z.; Gu, Q.; Zhang, L.; Poh, W.J.; Zhang, Z.; Wang, J.; Ng, H.Y. Effect of gradient profile in ceramic membranes on filtration characteristics: Implications for membrane development. *J. Membr. Sci.* **2020**, *595*, 117576. [CrossRef]

149. Murthy, S.K.; Sethu, P.; Vunjak-Novakovic, G.; Toner, M.; Radisic, M. Size-based microfluidic enrichment of neonatal rat cardiac cell populations. *Biomed. Microdevices* **2006**, *8*, 231–237. [CrossRef]

150. Ripperger, S.; Altmann, J. Crossflow microfiltration–state of the art. *Sep. Purif. Technol.* **2002**, *26*, 19–31. [CrossRef]

151. VanDelinder, V.; Groisman, A. Perfusion in microfluidic cross-flow: Separation of white blood cells from whole blood and exchange of medium in a continuous flow. *Anal. Chem.* **2007**, *79*, 2023–2030. [CrossRef]

152. Ji, H.M.; Samper, V.; Chen, Y.; Heng, C.K.; Lim, T.M.; Yobas, L. Silicon-based microfilters for whole blood cell separation. *Biomed. Microdevices* **2008**, *10*, 251–257. [CrossRef]

153. Chen, X.; Cui, D.; Zhang, L. Isolation of plasma from whole blood using a microfludic chip in a continuous cross-flow. *Chin. Sci. Bull.* **2009**, *54*, 324–327. [CrossRef]

154. Sethu, P.; Sin, A.; Toner, M. Microfluidic diffusive filter for apheresis (leukapheresis). *Lab Chip* **2006**, *6*, 83–89. [CrossRef]

155. Chen, X.; Cui, D.; Liu, C.; Li, H.; Chen, J. Continuous flow microfluidic device for cell separation, cell lysis and DNA purification. *Anal. Chim. Acta* **2007**, *584*, 237–243. [CrossRef]

156. Mielnik, M.M.; Ekatpure, R.P.; Sætran, L.R.; Schönfeld, F. Sinusoidal crossflow microfiltration device—Experimental and computational flowfield analysis. *Lab Chip* **2005**, *5*, 897–903. [CrossRef]

157. Moorthy, J.; Beebe, D.J. In situ fabricated porous filters for microsystems. *Lab Chip* **2003**, *3*, 62–66. [CrossRef]

158. Aran, K.; Fok, A.; Sasso, L.A.; Kamdar, N.; Guan, Y.; Sun, Q.; Ündar, A.; Zahn, J.D. Microfiltration platform for continuous blood plasma protein extraction from whole blood during cardiac surgery. *Lab Chip* **2011**, *11*, 2858–2868. [CrossRef]

159. Lo, M.; Zahn, J.D. Development of a multi-compartment microfiltration device for particle fractionation. In Proceedings of the 16th International Conference on Miniaturized Systems for Chemistry and Life Sciences, Okinawa, Japan, 28 October–1 November 2012.

160. Yamada, M.; Seki, M. Hydrodynamic filtration for on-chip particle concentration and classification utilizing microfluidics. *Lab Chip* **2005**, *5*, 1233–1239. [CrossRef]

161. Matsuda, M.; Yamada, M.; Seki, M. Blood cell classification utilizing hydrodynamic filtration. *Electron. Commun. Jpn.* **2011**, *94*, 1–6. [CrossRef]

162. Yamada, M.; Seki, M. Microfluidic particle sorter employing flow splitting and recombining. *Anal. Chem.* **2006**, *78*, 1357–1362. [CrossRef]

163. Chiu, Y.-Y.; Huang, C.-K.; Lu, Y.-W. Enhancement of microfluidic particle separation using cross-flow filters with hydrodynamic focusing. *Biomicrofluidics* **2016**, *10*, 11906. [CrossRef]

164. Yang, S.; Ündar, A.; Zahn, J.D. A microfluidic device for continuous, real time blood plasma separation. *Lab Chip* **2006**, *6*, 871–880. [CrossRef]

165. Kersaudy-Kerhoas, M.; Dhariwal, R.; Desmulliez, M.P.Y.; Jouvet, L. Hydrodynamic blood plasma separation in microfluidic channels. *Microfluid. Nanofluid.* **2010**, *8*, 105. [CrossRef]

166. Jäggi, R.D.; Sandoz, R.; Effenhauser, C.S. Microfluidic depletion of red blood cells from whole blood in high-aspect-ratio microchannels. *Microfluid. Nanofluid.* **2007**, *3*, 47–53. [CrossRef]

167. Wei Hou, H.; Gan, H.Y.; Bhagat, A.A.S.; Li, L.D.; Lim, C.T.; Han, J. A microfluidics approach towards high-throughput pathogen removal from blood using margination. *Biomicrofluidics* **2012**, *6*, 24115. [CrossRef]

168. Fåhraeus, R. The suspension stability of the blood. *Physiol. Rev.* **1929**, *9*, 241–274. [CrossRef]

169. Geng, Z.; Zhang, L.; Ju, Y.; Wang, W.; Li, Z. A plasma separation device based on centrifugal effect and Zweifach-Fung effect. In Proceedings of the 15th International Conference on Miniaturized Systems for Chemistry and Life Sciences, Seattle, WA, USA, 2–6 October 2011; pp. 224–226.

170. Dong, T.; Yang, Z.; Su, Q.; Tran, N.M.; Egeland, E.B.; Karlsen, F.; Zhang, Y.; Kapiris, M.J.; Jakobsen, H. Integratable non-clogging microconcentrator based on counter-flow principle for continuous enrichment of CaSki cells sample. *Microfluid. Nanofluid.* **2011**, *10*, 855–865. [CrossRef]

171. Hønsvall, B.K.; Altin, D.; Robertson, L.J. Continuous harvesting of microalgae by new microfluidic technology for particle separation. *Bioresour. Technol.* **2016**, *200*, 360–365. [CrossRef] [PubMed]

172. Mossige, E.J.; Jensen, A.; Mielnik, M.M. An experimental characterization of a tunable separation device. *Microfluid. Nanofluid.* **2016**, *20*, 1–10. [CrossRef]

173. Mossige, E.J.; Jensen, A.; Mielnik, M.M. Separation and concentration without clogging using a high-throughput tunable filter. *Phys. Rev. Appl.* **2018**, *9*, 54007. [CrossRef]

174. Mossige, E.J.; Edvardsen, B.; Jensen, A.; Mielnik, M.M. A tunable, microfluidic filter for clog-free concentration and separation of complex algal cells. *Microfluid. Nanofluid.* **2019**, *23*, 56. [CrossRef]

175. Hsu, C.-H.; Di Carlo, D.; Chen, C.; Irimia, D.; Toner, M. Microvortex for focusing, guiding and sorting of particles. *Lab Chip* **2008**, *8*, 2128–2134. [CrossRef]

176. Bhardwaj, P.; Bagdi, P.; Sen, A.K. Microfluidic device based on a micro-hydrocyclone for particle–liquid separation. *Lab Chip* **2011**, *11*, 4012–4021. [CrossRef]

177. Chand, R.; Ramalingam, S.; Neethirajan, S. A 2D tran sition-metal dichalcogenide MoS2 based novel nanocomposite and nanocarrier for multiplex miRNA detection. *Nanoscale* **2018**, *10*, 8217–8225. [CrossRef]

178. Martins, G.V.; Marques, A.C.; Fortunato, E.; Sales, M.G.F. Wax-printed paper-based device for direct electrochemical detection of 3-nitrotyrosine. *Electrochim. Acta* **2018**, *284*, 60–68. [CrossRef]

179. Reich, P.; Preuss, J.A.; Bahner, N.; Bahnemann, J. Impedimetric aptamer-based biosensors: Principles and techniques. *Adv. Biochem. Eng. Biotechnol.* **2020**, *174*, 17–41.

180. Preuss, J.A.; Reich, P.; Bahner, N.; Bahnemann, J. Impedimetric aptamer-based biosensors: Applications. *Adv. Biochem. Eng. Biotechnol.* **2020**, *174*, 43–91.

181. Mousavi, M.P.S.; Ainla, A.; Tan, E.K.W.; Abd El-Rahman, M.K.; Yoshida, Y.; Yuan, L.; Sigurslid, H.H.; Arkan, N.; Yip, M.C.; Abrahamsson, C.K.; et al. Ion sensing with thread-based potentiometric electrodes. *Lab Chip* **2018**, *18*, 2279–2290. [CrossRef]

182. Shiddiky, M.J.A.; Park, H.; Shim, Y.-B. Direct Analysis of Trace Phenolics with a Microchip: In-Channel Sample Preconcentration, Separation, and Electrochemical Detection. *Anal. Chem.* **2006**, *78*, 6809–6817. [CrossRef] [PubMed]

183. Hiraiwa, M.; Kim, J.H.; Lee, H.B.; Inoue, S.; Becker, A.L.; Weigel, K.M.; Cangelosi, G.A.; Lee, K.H.; Chung, J.H. Amperometric immunosensor for rapid detection of *Mycobacterium tuberculosis*. *J. Micromech. Microeng.* **2015**, *25*, 055013. [CrossRef] [PubMed]

184. Elshafey, R.; Tlili, C.; Abulrob, A.; Tavares, A.C.; Zourob, M. Label-free impedimetric immunosensor for ultrasensitive detection of cancer marker Murine double minute 2 in brain tissue. *Biosens. Bioelectron.* **2013**, *39*, 220–225. [CrossRef] [PubMed]

185. Cecchetto, J.; Carvalho, F.C.; Santos, A.; Fernandes, F.C.B.; Bueno, P.R. An impedimetric biosensor to test neat serum for dengue diagnosis. *Sens. Actuators B Chem.* **2015**, *213*, 150–154. [CrossRef]

186. Ding, J.W.; Qin, W. Recent advances in potentiometric biosensors. *TrAC Trends Anal. Chem.* **2020**, *124*, 115803. [CrossRef]

187. Luo, X.L.; Davis, J.J. Electrical biosensors and the label free detection of protein disease biomarkers. *Chem. Soc. Rev.* **2013**, 5944–5962. [CrossRef] [PubMed]

188. Lan, W.-J.; Zou, X.U.; Hamedi, M.M.; Hu, J.; Parolo, C.; Maxwell, E.J.; Bühlmann, P.; Whitesides, G.M. Paper-Based Potentiometric Ion Sensing. *Anal. Chem.* **2014**, *86*, 9548–9553. [CrossRef]

189. Han, Y.X.; Chen, J.; Li, Z.; Chen, H.L.; Qiu, H.D. Recent progress and prospects of alkaline phosphatase biosensor based on fluorescence strategy. *Biosens. Bioelectron.* **2020**, *148*, 111811. [CrossRef] [PubMed]

190. Wang, Y.; Li, Z.H.; Wang, J.; Li, J.H.; Lin, Y.H. Graphene and graphene oxide: Biofunctionalization and applications in biotechnology. *Trends Biotechnol.* **2011**, *29*, 205–212. [CrossRef]

191. Takemura, K.; Adegoke, O.; Suzuki, T.; Park, E.Y. A localized surface plasmon resonance-amplified immunofluorescence biosensor for ultrasensitive and rapid detection of nonstructural protein 1 of Zika virus. *PLoS ONE* **2019**, *14*, e0211517. [CrossRef]

192. Sieben, V.J.; Floquet, C.F.A.; Ogilvie, I.R.G.; Mowlem, M.C.; Morgan, H. Microfluidic colourimetric chemical analysis system: Application to nitrite detection. *Anal. Methods* **2010**, *2*, 484–491. [CrossRef]

193. Wang, X.; Qian, X.; Beitler, J.J.; Chen, Z.G.; Khuri, F.R.; Lewis, M.M.; Shin, H.J.C.; Nie, S.; Shin, D.M. Detection of Circulating Tumor Cells in Human Peripheral Blood Using Surface-Enhanced Raman Scattering Nanoparticles. *Cancer Res.* **2011**, *71*, 1526–1532. [CrossRef] [PubMed]

194. Wu, X.; Xia, Y.; Huang, Y.; Li, J.; Ruan, H.; Luo, L.; Yang, S.; Shen, Z.; Wu, A. Improved SERS-Active Nanoparticles with Various Shapes for CTC Detection without Enrichment Process with Supersensitivity and High Specificity. *ACS Appl. Mater. Interfaces* **2016**, *8*, 19928–19938. [CrossRef]

195. Quang, L.X.; Lim, C.; Seong, G.H.; Choo, J.; Do, K.J.; Yoo, S.-K. A portable surface-enhanced Raman scattering sensor integrated with a lab-on-a-chip for field analysis. *Lab Chip* **2008**, *8*, 2214–2219. [CrossRef]

196. Pashchenko, O.; Shelby, T.; Banerjee, T.; Santra, S. A comparison of optical, electrochemical, magnetic, and colorimetric point-of-care biosensors for infectious disease diagnosis. *ACS Infect. Dis.* **2018**, *4*, 1162–1178. [CrossRef]

197. Schotter, J.; Kamp, P.B.; Becker, A.; Puhler, A.; Reiss, G.; Bruckl, H. Comparison of a prototype magnetoresistive biosensor to standard fluorescent DNA detection. *Biosens. Bioelectron.* **2004**, *19*, 1149–1156. [CrossRef] [PubMed]

198. Santiesteban, O.J.; Kaittanis, C.; Perez, J.M. Identification of toxin inhibitors using a magnetic nanosensor-based assay. *Small* **2014**, *10*, 1202–1211. [CrossRef] [PubMed]

199. Sideris, C.; Khial, P.P.; Hajimiri, A. Design and implementation of reference-free drift-cancelling CMOS magnetic sensors for biosensing applications. *IEEE J. Solid-State Circuits* **2018**, *53*, 3065–3075. [CrossRef]

200. Hong, S.L.; Zhang, N.; Qin, L.; Tang, M.; Ai, Z.; Chen, A.; Wang, S.; Liu, K. An automated detection of influenza virus based on 3-D magnetophoretic separation and magnetic label. *Analyst* **2021**, *146*, 930–936. [CrossRef] [PubMed]

201. Wu, K.; Klein, T.; Krishna, V.D.; Su, D.Q.; Perez, A.M.; Wang, J.P. Portable GMR handheld platform for the detection of influenza A virus. *ACS Sens.* **2017**, 21594–21601. [CrossRef] [PubMed]

202. Wu, K.; Liu, J.M.; Saha, R.; Su, D.Q.; Krishna, V.D.; Cheeran, M.C.J.; Wang, J.P. Magnetic particle spectroscopy for detection of influenza A virus subtype H1N1. *ACS Appl. Mater. Interfaces* **2020**, *12*, 13686–13697. [CrossRef]

203. Wang, C.; Liu, M.; Wang, Z.; Li, S.; Deng, Y.; He, N. Point-of-care diagnostics for infectious diseases: From methods to devices. *Nano Today* **2021**, *37*, 101092. [CrossRef]

204. Wongkaew, N.; Simsek, M.; Griesche, C.; Baeumner, A.J. Functional nanomaterials and nanostructures enhancing electrochemical biosensors and lab-on-a-chip performances: Recent progress, applications, and future perspective. *Chem. Rev.* **2019**, *119*, 120–194. [CrossRef]

205. Jiang, P.J.; Guo, Z.J. Fluorescent detection of zinc in biological systems: Recent development on the design of chemosensors and biosensors. *Coord. Chem. Rev.* **2004**, *248*, 205–229. [CrossRef]

206. Xianyu, Y.L.; Wang, Q.L.; Chen, Y.P. Magnetic particles-enabled biosensors for point-of-care testing. *TrAC Trends Anal. Chem.* **2018**, *106*, 213–224. [CrossRef]

207. Leong, W.; Wang, D.-A. Cell-laden polymeric microspheres for biomedical applications. *Trends Biotechnol.* **2015**, *33*, 653–666. [CrossRef]

208. Le, T.T.; Andreadakis, Z.; Kumar, A.; Román, R.G.; Tollefsen, S.; Saville, M.; Mayhew, S. The COVID-19 vaccine development landscape. *Nat. Rev. Drug Discov.* **2020**, *19*, 305–306. [CrossRef]

209. Van Riel, D.; de Wit, E. Next-generation vaccine platforms for COVID-19. *Nat. Mater.* **2020**, *19*, 810–812. [CrossRef]

210. Smith, T.R.F.; Patel, A.; Ramos, S.; Elwood, D.; Zhu, X.; Yan, J.; Gary, E.N.; Walker, S.N.; Schultheis, K.; Purwar, M. Immunogenicity of a DNA vaccine candidate for COVID-19. *Nat. Commun.* **2020**, *11*, 2601. [CrossRef]

211. Khademhosseini, A.; Langer, R. Microengineered hydrogels for tissue engineering. *Biomaterials* **2007**, *28*, 5087–5092. [CrossRef]

212. Rytting, E.; Nguyen, J.; Wang, X.; Kissel, T. Biodegradable polymeric nanocarriers for pulmonary drug delivery. *Expert Opin. Drug Deliv.* **2008**, *5*, 629–639. [CrossRef]

MDPI

St. Alban-Anlage 66

4052 Basel

Switzerland

Tel. +41 61 683 77 34

Fax +41 61 302 89 18

www.mdpi.com

Micromachines Editorial Office

E-mail: micromachines@mdpi.com

www.mdpi.com/journal/micromachines